Active Matrix Liquid Crystal Displays

Active Matrix Liquid Crystal Displays

Willem den Boer

ELSEVIER

AMSTERDAM • BOSTON • HEIDELBERG • LONDON
NEW YORK • OXFORD • PARIS • SAN DIEGO
SAN FRANCISCO • SINGAPORE • SYDNEY • TOKYO

Newnes is an imprint of Elsevier

Newnes

Newnes is an imprint of Elsevier
30 Corporate Drive, Suite 400, Burlington, MA 01803, USA
Linacre House, Jordan Hill, Oxford OX2 8DP, UK

Recognizing the importance of preserving what has been written, Elsevier prints its books on acid-free
paper whenever possible.

Library of Congress Cataloging-in-Publication Data
Application submitted.

British Library Cataloguing-in-Publication Data
A catalogue record for this book is available from the British Library.

ISBN-10: 0-7506-7813-5
ISBN-13: 978-0-7506-7813-1

For information on all Newnes publications
visit our Web site at www.books.elsevier.com

Transferred to Digital Printing in 2009

To Deirdre

Contents

Preface

The meteoric rise of the active matrix LCD over less than two decades to undisputed dominance as a flat panel display has been breathtaking. The technology behind this remarkable progress will be summarized in this book. Manufacturing of AMLCDs is a more than $40 billion industry and now plays an important role in the economy of several Asian countries. This book will address the fundamentals of LCD operation and the principles of active matrix addressing. The reader will become familiar with the construction and manufacturing methods of AMLCDs, as well as with drive methods; performance characteristics; recent improvements in image quality; and applications in cellular phones, portable computers, desktop monitors, and LCD televisions.

This book is on an introductory level and intended for students, engineers, managers, educators, IP lawyers, research scientists, and technical professionals. Emphasis has been placed on explaining underlying principles in the simplest possible way without relying extensively on equations. Last, but not least, the book is intended for those inquiring minds who, for many hours every day, look at the LCD displays of notebook computers, flat panel monitors, and televisions and simply wonder how they work.

This book grew out of the course materials from seminars I presented several times at UCLA, at the gracious invitation of course organizer Larry Tannas. The course materials were further updated when the Society for Information Display kindly invited me to present Short Courses on AMLCDs at the annual SID Symposiums in 2002 and 2003.

A book like this would not have been possible without frequent interaction and discussions with colleagues in the field. Working with them has always been a great pleasure and I would like to mention specifically some of my former coworkers at OIS Optical Imaging Systems, Inc. and at Planar Systems, Inc. They include Adi Abileah, Steve Aggas, Tom Baker, Yair Baron, Bill Barrow, Mike Boyd, Young Byun, Vin Cannella, Mark Friends, Pat Green, Tieer Gu, Chris King, Terrance Larsson, Alan Lien, Darrin Lowe, Yiwei Lu, Fan Luo, Tin Nguyen, Cheng-bin Qiu, Scott Robinson, Scott Smith, Scott Thomsen, Dick Tuenge, Victor Veerasamy, Mimi Wang, Moshe Yang, Mei Yen, Zvi Yaniv, and John Zhong.

Finally, I would like to thank the Elsevier Science organization for the opportunity to write this book and for their support during its production.

Willem den Boer

1

Introduction

1.1 Historical Perspective

Electronic displays have, for many years, been the window to the world in television as well as the primary human interface to computers. In today's information society, they play an increasingly important and indispensable role in communication, computing, and entertainment devices.

The venerable cathode ray tube (CRT) has been around for more than 100 years and has been the workhorse for television displays and, until recently, for computer screens. As one of the few surviving electronic devices based on vacuum tubes, the CRT can boast an unrivaled success as a low-cost color display with good image quality. Its large depth, weight, and power consumption, however, have limited its use to nonportable applications.

From the early days of electronic display development, a flat panel display was considered a very attractive alternative to the bulky CRT. For decades, display engineers searched for flat panel display technologies to replace the CRT in many applications. In spite of many attempts to develop flat CRTs, plasma displays, and other low-profile displays, commercial success remained elusive for many years. Finally, by the 1990s several technologies were making significant inroads to achieve this goal. In particular, active matrix liquid crystal displays (AMLCDs) and plasma displays demonstrated large sizes and high image quality comparable to CRTs.

The success of AMLCDs, the subject of this book, is the culmination of two significant developments: liquid crystal cell technology and large-area microelectronics on glass. For more than two decades these technologies have been refined and an extensive infrastructure for manufacturing equipment and materials has been established, especially in Asia.

Liquid crystals were discovered in 1888 by the Austrian botanist Friedrich Reinitzer. He experimented with the cholesterol-type organic fluid cholesteryl benzoate and found that, upon heating, it underwent a phase transition from a milky fluid to a mostly transparent fluid. This was later explained as a transition from an optically and electrically anisotropic fluid to an isotropic fluid. The anisotropy (i.e., the difference in dielectric constant and refractive index for different orientations of the molecules in the fluid) led to the analogy with the anisotropy of solid crystals, hence the name liquid crystal. The technological and commercial potential of liquid crystal was not realized until the 1960s, when RCA developed the first liquid crystal displays (LCDs) based on the dynamic scattering effect [1]. The twisted nematic (TN) mode of operation, on which many current LCDs are based, was first described by Schadt and Helfrich [2] in 1971 and, independently around the same time, by Fergason [3]. Twisted nematic LCDs appeared on the scene in the early 1970s in electronic wrist watches and in calculators.

LCDs quickly dominated in small portable applications due to the compatibility of the simple reflective-type LCD with low-power CMOS driving circuitry and therefore with battery operation. In addition, high-volume manufacturing led to very low cost. The market for small, direct-driven, segmented, TN LCDs in portable devices increased rapidly during the 1970s. They were initially mostly used in a reflective mode, relying on ambient light for legibility. Since each segment in a direct-driven alphanumeric display needs to be separately connected to the control electronics, the information content of this type of display is very limited, usually to one or two lines of text or numbers. Other drawbacks of the reflective mode included the difficulty of implementing color, the dependence on ambient lighting, and the parallax caused by the separation of about 0.5–1 mm between the back reflector and the LC layer.

The mass market for electronic wrist watches, calculators, and other applications, however, allowed investments in manufacturing and further development.

For displays with higher information content, the large number of picture elements (pixels) precluded the individual addressing of every pixel. This led to the development of matrix addressing in which an array of $M \times N$ pixels is addressed by applying pulses to each of its M rows and N columns. It reduces the number of interconnects to the external addressing circuitry from $M \times N$ to $M + N$. For example, a 100×100-pixel display now required $100 + 100 = 200$ interconnections instead of $100 \times 100 = 10,000$. Such passive matrix displays are usually operated with a one-line-at-a-time addressing method called multiplexing. The TN LCD was limited to only about 10 rows of pixels because of its gradual transmission-voltage curve. Many improvements in passive matrix addressing have been proposed and implemented over the last fifteen years, notably the super-twisted nematic (STN) LCD. However, their performance in terms of viewing angle, response time, gray scale, and contrast ratio has generally fallen short of what was possible with a single direct-driven pixel.

Early during the development of LCD technology, the limitations of direct multiplexing or passive matrix addressing were recognized. A solution was proposed by Lechner et al. [4] and by Marlowe and Nester [5]. By incorporating a switch at each pixel in a matrix display, the voltage across each pixel could be controlled independently. The same high-contrast ratio of more than 200, obtained in simple, direct-driven backlit displays, could then also be achieved in high-information-content displays.

Peter Brody and coworkers [6] constructed the first so-called active matrix LCDs (AMLCDs) with CdSe thin film transistors (TFTs) as the switching elements. The TFTs in the array act merely as ON/OFF switches and do not have an amplifying function. For an electronic engineer, the term "active matrix" may therefore be inappropriate. It is now, however, commonly used and its definition has been extended to include arrays of switching elements other than TFTs, such as diodes.

The CdSe TFTs used for the first AMLCDs turned out to be a temporary solution. Semiconductors such as CdSe are not compatible with standard processing in the microelectronic industry, which uses mainly silicon as the semiconductor material. Advanced photolithographic and etching processes have been developed over the years for silicon devices and this technology is not readily applicable to CdSe TFTs. Dr. Brody has, nonetheless, remained a strong and vocal proponent of CdSe-based AMLCDs over the years, even in the face of the overwhelming success of silicon thin film-based TFT LCDs.

Polycrystalline Si materials and devices are more familiar to semiconductor process engineers and were developed for use in AMLCDs in the early 1980s. The first LC pocket television marketed by Seiko Epson in 1983 used a poly-Si TFT active matrix [7] and was the very first commercial application of AMLCDs. The early poly-Si TFTs required high-temperature processing and therefore used expensive quartz substrates.

A color LCD was obtained by subdividing the pixel into three subpixels with red, green, and blue color filters. Since the color filters absorb a large portion of the light, these color LCDs required a backlight to operate in a transmissive rather than a reflective mode to be useful in most ambient lighting conditions.

In parallel to the early development of LC cell technology and CdSe TFTs, thin film amorphous silicon (a-Si) was investigated in the 1970s. The rationale behind this interest was initially not its potential use for LCDs, but rather its promise for low-cost solar cells. A major development occurred at the University of Dundee in Scotland in 1979, when LeComber et al [8] developed the first TFT with a-Si as the semiconductor material and suggested the active matrix LCD as one of its applications. Interestingly, a patent on the basic a-Si TFT was never filed, since this work was performed at an academic institution. The University of Dundee had been a pioneer in the development and understanding of amorphous silicon materials under the leadership of Professor W.E. Spear and Dr. P.G. LeComber. Unfortunately, Dr. LeComber witnessed only the start of the tremendous growth in AMLCDs based on a-Si TFTs; he died at the age of 51 of a fatal heart attack while vacationing in Switzerland in 1992.

Amorphous silicon can be deposited on low-cost, large-area glass substrates at a temperature below 300°C. It was more attractive than the early polycrystalline technology for active matrix LCDs, which needed much higher process temperatures and more process steps. A pocket television with a-Si TFTs was put on the market by Matsushita in 1984 [9].

Soon after the first TFT LCD with amorphous silicon TFTs was introduced, the a-Si TFT overshadowed poly-Si TFTs as the semiconductor device of choice for AMLCDs. The first TFT LCDs had a diagonal size of 2–3 in. and were mainly used in small portable televisions. The performance of the color AMLCD was initially improved for high-end applications such as aircraft cockpit displays. In avionics, cost was a secondary concern and emphasis was placed on the highest possible performance in terms of legibility under any lighting conditions.

While a market was established, volume production capability was gradually increased at several Japanese companies. This was accompanied by a scale-up in glass substrate size from wafer-like sizes to around 300×400 mm. A significant breakthrough occurred in the late 1980s, when the first laptop computers with 10-in. diagonal size TFT LCDs were marketed by several companies, including IBM, NEC, Sharp, Toshiba, and Hitachi. This scale-up was made possible by the gradual increase in manufacturing yields by process improvement methods borrowed from the semiconductor industry. Laptop and notebook computers turned out to be the killer application for active matrix liquid crystal displays. They addressed the need of traveling businessmen for lightweight computers with high image quality, and thin and low-power flat panel displays.

By 1996 the manufacturing substrate size had grown to 550×650 mm and many improvements in processing and materials were implemented. In the mid-1990s Korean companies started mass production of AMLCD modules, followed a few years later by massive investments by several companies in Taiwan.

The late 1990s also witnessed a revival of poly-Si TFT LCDs for small displays. Several companies succeeded in producing low-temperature poly-Si TFTs processed at temperatures below 600°C, compatible with lower-cost glass. The main attraction of poly-Si TFTs is their higher current-carrying capability, allowing the integration of some of the drive electronics on the glass.

Application of a-Si TFT LCDs in notebook computers facilitated large investments in their manufacturing infrastructure. This, in turn, made possible the introduction of space- and power-saving, flat-panel desktop monitors based on AMLCD panels with improved viewing angles.

Several methods to improve the viewing angle, including compensation films and different LC modes (in-plane-switching and multi-domain vertical alignment) were introduced after 1995 and implemented in 17-in. and larger monitors. In 2003 TFT LCDs surpassed CRTs in terms of revenue for computer monitors.

Recent further improvements in brightness, color performance, viewing angle, and response time have led to the development of LCD television with superior image quality and progressively larger screens, now well beyond 40 in. in diagonal size. LCD television is the final frontier for the AMLCD and a number of companies have started production on glass substrates with $1-4$ m^2 size to participate in this rapidly growing market.

Another application where AMLCDs have attained a large market share is handheld devices (i.e., in PDAs, digital cameras, camcorders, and mobile phones). With the introduction of 3G cellular phone service and built-in cameras, the demand for high-contrast, video-rate color displays have allowed the AMLCD to replace many of the poorer-performing passive matrix LCDs.

Parallel to the development of direct-view displays, microdisplays for projection have been developed since the late 1970s. It was realized early on that the LCD is a light valve that can act as an electronically controlled slide in a slide projector. Business-grade front projectors appeared on the scene in the early 1990s with three high-resolution poly-Si TFT LCDs. They are now commonplace in many meeting rooms and classrooms across the world and can weigh less than 3 lbs. Rear-projection, high-definition television based on reflective microdisplays (liquid crystal on silicon [LCOS]) have entered the marketplace as well.

Microdisplays are also used in personal viewers such as viewfinders for digital cameras and camcorders.

The success of AMLCD technology is the result of many years of close cooperation among scientists and engineers from different disciplines. They include organic chemists, physicists, optical, electrical, electronic, mechanical, packaging, and manufacturing engineers, all supported by increasing revenues from sales of LCDs.

Figure 1.1 illustrates the exponential increase in the total market for AMLCDs. The development shows a remarkable parallel with the semiconductor industry in the 1980s and 1990s, also indicated in the plot. Some fluctuations in the LCD revenue curve can be attributed to supply-demand cycles. Based on an extrapolation of this plot, the AMLCD industry will achieve $100 billion in annual revenues around the year 2010.

The market for AMLCDs is usually subdivided into small displays with a diagonal size of less than 10 in. (in PDAs, cell phones, digital cameras, camcorders, etc.), medium displays of 10–20 in. (in notebook computers and desktop monitors), and large displays exceeding 20 in. (mostly in televisions). Figure 1.2 shows the recent rapid unit growth in all applications. Although the small displays have the largest unit sales, the larger displays obviously represent a large part of the total revenues.

AMLCDs are also increasingly used in medical, industrial, and retail applications, often with a touch panel included.

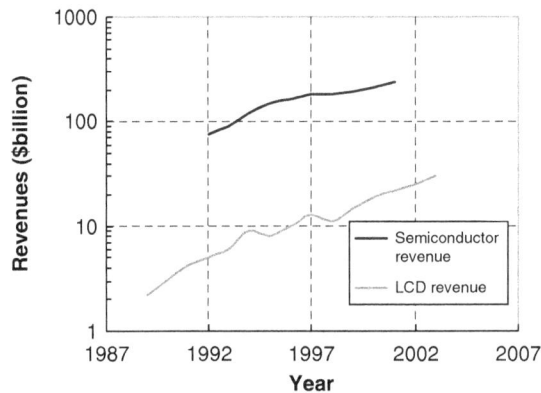

Figure 1.1: Growth of the semiconductor IC and LCD market.

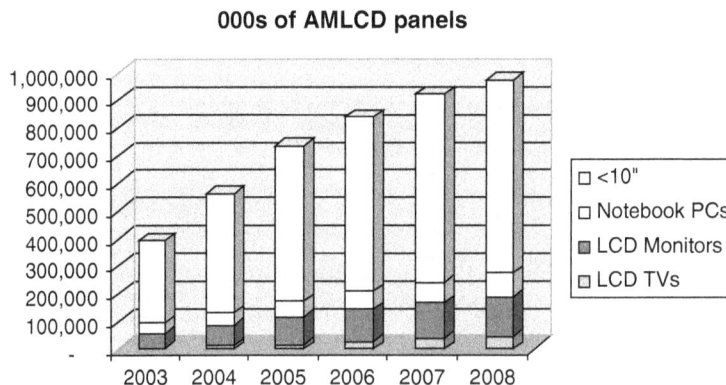

Figure 1.2: Market forecast for TFT LCDs (courtesy of Displaysearch).

The AMLCD manufacturers are supported by a large infrastructure of equipment and material suppliers, which continue to improve efficiency and reduce cost. They include color filter, polarizer, and optical film manufacturers, driver IC and controller IC vendors, packaging firms, backlight manufacturers, and suppliers of materials such as LC fluids and alignment layers.

The momentum the LCD industry has gained is difficult to supplant by alternative flat panel display technologies, even if, on paper, they may have advantages over LCDs.

1.2 Liquid Crystal Properties

After their discovery by Reinitzer in 1888, liquid crystals were, for many decades, considered interesting only from an academic point of view. Liquid crystals are an intermediate phase between crystalline solids and isotropic liquids, and combine certain characteristic properties of the crystal structure with those of a deformable fluid. Display devices utilize both their fluidity and the anisotropy associated with their crystalline character. The anisotropy causes the dielectric constant and refractive index of the LC fluid to depend on the orientation of its molecules.

Nematic liquid crystals are commercially the most interesting type. Upon heating, most crystalline solids undergo a phase transition to an isotropic liquid. The nematic intermediate phase (mesophase) occurs in certain, mostly organic, substances in a temperature range between the solid and isotropic liquid state. Figure 1.3 shows an example of the molecular structure of a liquid crystal, p-methoxybenzylidene-p-n-butylaniline (MBBA), with its nematic temperature range.

The liquid crystal (LC) molecules are generally elongated in shape and have a length of around 2 nm. Because of their "cigar" shape they tend to line up more or less parallel to each other in the lowest energy state. The average orientation axis along the molecules is a unit vector **n**, called the director. Nematic LC molecules are not polar, so there is no differentiation between **n** and −**n**. The dielectric constant and refractive index of the LC is different along the director and perpendicular to the director, as shown in Fig. 1.4, giving rise to dielectric and optical anisotropy, respectively. The dielectric anisotropy

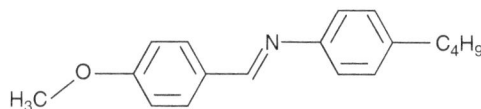

Figure 1.3: Example of molecular structure of LC – MBBA, with a nematic range of 21–48°C.

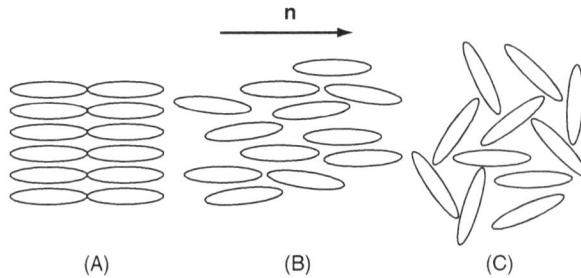

Figure 1.4: Orientation of LC molecules in (A) smectic, (B) nematic, and (C) isotropic phase.

makes it possible to change the orientation of the LC molecules in an electric field, crucial for application in electro-optical devices. The optical anisotropy leads to birefringence effects (described later in this chapter) and is essential for the modulation of polarized light in display operation.

In bulk, liquid crystal tends to form microdroplets. Within the droplets there is one director orientation, but the director can be different for adjacent droplets. Although most high-purity liquid crystals are transparent, this explains the milky appearance of bulk LC, as scattering of light occurs at the boundaries of the microdroplets.

When nematic LC is heated beyond a certain temperature, a phase transition occurs to an isotropic liquid. The nematic-isotropic transition temperature is often referred to as the clearing point or clearing temperature because of the drastic reduction in light scattering that occurs when the fluid no longer consists of microdroplets with different directors. Just below the clearing point, the optical and dielectric anisotropy of the LC fluid starts to decline, until at the clearing point there is a single dielectric constant and refractive index in the isotropic state. Above the clearing point, the LC fluid no longer has the desirable optical and dielectric anisotropy, and display operation fails.

When the temperature is lowered, the LC fluid undergoes another phase transition from the nematic to the smectic phase. In the smectic phase, the LC molecules form a layered structure and obtain a higher viscosity to become more grease-like. When approaching the smectic phase, the LC response to an electric field change becomes very sluggish, and below the transition temperature the display operation fails.

A large variety of nematic liquid crystals has been synthesized over the years, each with its own molecular formula and nematic temperature range. For MBBA in the example of Fig. 1.3, the nematic range is 21–48°C. This range is too limited for practical applications. Commercial fluids generally consist of a mixture of two or more liquid crystal components and have a much wider nematic range.

Another LC type is the cholesteric (chiral nematic) phase in which the molecular interaction causes a helical twist of the director orientation **n**, as illustrated in Fig. 1.5. The pitch (p) of the cholesteric phase is defined as the distance over which the director twists 360 degrees. In display devices based on the twisted nematic effect, small concentrations of cholesteric material are usually added to produce a preferred twist sense in the cell and avoid display artifacts.

Nematic liquid crystal is a viscous fluid with a viscosity at room temperature about ten times higher than that of water. The viscosity plays an important role in the response time of LC fluids to electric fields, and a low viscosity is required for displays operating at video rates.

The lowest energy state for bulk nematic liquid crystal occurs when there is a single director throughout the fluid (i.e., when all molecules have the same orientation). The bulk elastic properties associated with the curvature of the director **n** can be described by three elastic constants. These constants K_{11}, K_{22}, and K_{33} quantify the restoring torque opposing splay, twist, and bend deformation of the director orientation, respectively (Fig. 1.6).

The constants K_{11}, K_{22}, and K_{33} are of the order of 10^{-11} N and orders of magnitude smaller than the elastic constants in a solid-state material. This allows the LC fluid to readily deform when subjected to an electric field, which is essential for display operation. With increasing temperature, K_{11}, K_{22}, and K_{33} decrease as elasticity weakens.

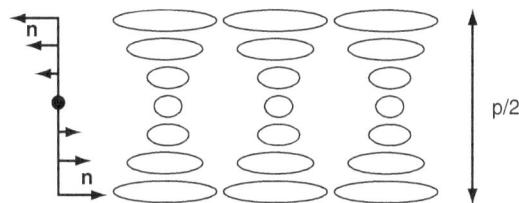

Figure 1.5: Cholesteric or chiral nematic phase.

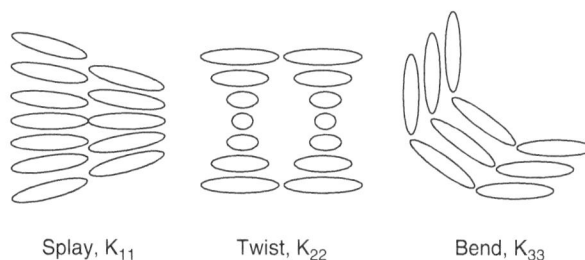

Splay, K_{11} Twist, K_{22} Bend, K_{33}

Figure 1.6: Elastic deformations in LC fluid and their associated elastic constants.

In Table 1.1 selected properties of a typical nematic liquid crystal fluid are shown. $V_{th,10}$ is the voltage at which 10% of transmittance change is obtained in display operation.

Table 1.1: Selected properties of a typical nematic liquid crystal mixture for display applications

Parameter	Symbol	Typical Value	Comments
Clearing point	T_{Clp}	80°C	Max. operating temperature
Smectic-Nematic transition	T_{S-N}	−40°C	Min. operating temperature
Optical anisotropy	$\Delta n = n_{\parallel} - n_{\perp}$	0.085=1.562−1.477	Determines optical behavior
Dielectric anisotropy	$\Delta\varepsilon = \varepsilon_{\parallel} - \varepsilon_{\perp}$	7=10.5−3.5	Determines behavior in electric field
Threshold voltage TN cell	$V_{th\,10}$	1.6 V	Voltage at 10% transmission in NB mode
Elastic constants	K_{11}, K_{22}, K_{33}	10^{-11} Newton	Important for response time
Rotational viscosity at 20°C	γ_1	100 mPa-s	Important for response time

For display applications LC is sandwiched in a thin layer of 2–10 µm between two substrates, usually glass, which have conducting electrodes on their inner surfaces. Near the substrate surface the LC molecules can exhibit alignment phenomena, which can be strengthened by depositing certain organic or inorganic films on the surface and treating the surface of the film to obtain a preferred orientation of the molecules. These so-called orienting layers or alignment layers force the director to assume a single orientation at and near the entire surface. The degree of anchoring to the surface is called the anchoring strength and depends on the orienting layer, its surface treatment, and the liquid crystal.

Examples of surface alignment are shown in Fig. 1.7. Orientation perpendicular to the surface is called homeotropic alignment and anchoring of the director parallel to the surface is called homogeneous alignment. Both types of alignment are used in practical displays.

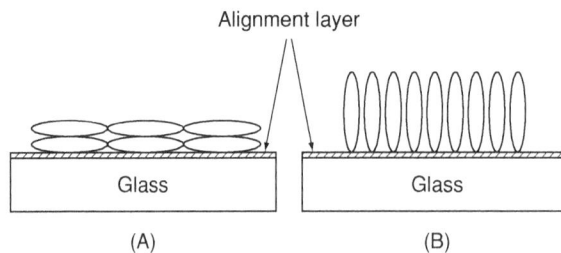

Figure 1.7: Homogeneous (A) and homeotropic (B) surface alignment.

Another important property of LC fluid, in particular for active matrix displays, is its resistivity. The LC pixel acts like a capacitance that stores the data voltage loaded once every frame time. If the pixel capacitance is leaky as a result of low resistivity of the LC fluid, the data voltage will decay during the frame time. It is then not possible to obtain a uniform, consistent, gray scale. A resistivity of 10^{11} Ωcm or higher is typically required for active matrix LCDs. The LC material therefore needs to be very pure and not contain moisture and ionic contamination, which can reduce its resistivity.

1.3 Polarization, Dichroism, and Birefringence

Most LCDs rely on the manipulation of polarized light to generate an image based on data from an electronic signal source. Before discussing LC modes of operation, it is therefore useful to address the basics of polarized light and of the optical anisotropy or birefringence allowing this manipulation.

Natural light consists of transverse electromagnetic waves which vibrate in arbitrary planes and can be described as the sum of two orthogonal waves. A polarizer (P) transmits only one of the two composing waves in one plane of vibration to obtain linearly polarized light (see Fig. 1.8). The intensity (or luminance) of the transmitted light is not more than 50% of that of the natural incident light. When a second polarizer (A, the analyzer) is placed behind P such that its plane of vibration is at an angle ϕ with that of P, the amplitude E_t of the wave transmitted through A is

$$E_t = E_i \cos\phi, \qquad (1.1)$$

where E_i is the amplitude of the incident wave on A.

Since the light intensity is proportional to the square of the amplitude, the transmitted light intensity is

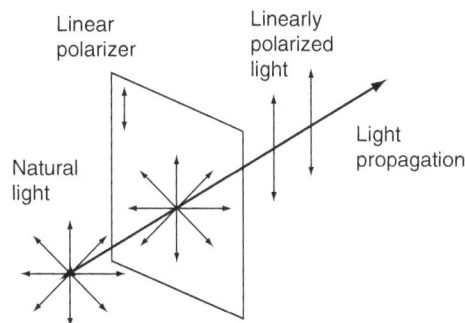

Figure 1.8: Linear polarization of light.

$$T = T_0 \cos^2 \phi, \qquad (1.2)$$

where T_0 is the intensity of transmitted light through the first polarizer.

For crossed ideal polarizers, $\phi = 90°$ and transmission is zero. For parallel ideal polarizers, $\phi = 0°$ and transmission is 50%.

Practical polarizers do not follow Eqs. 1.1 and 1.2 exactly. Polarizers used in LCDs usually consist of stretched polymer films, such as polyvinylacetate (PVA), doped with iodine or other specific additives. The stretching process makes the polymer optically anisotropic. The iodine doping causes strong absorption of incident light along the optical axis, so that only one linearly polarized component of the incident light is transmitted.

The optical absorption along the optical axis is referred to as dichroism.

Polarizers are characterized by their transmittance and degree of polarization (also called polarization or polarizing efficiency).

The polarization efficiency *Eff* is defined as

$$Eff = \sqrt{\frac{T_1 - T_2}{T_1 + T_2}}, \qquad (1.3)$$

where T_1 is the transmission of two parallel polarizers and T_2 the transmission of two crossed polarizers.

For LCDs using two polarizers, one outside each of its glass substrates, the maximum contrast ratio CR_{max} is limited by the polarization efficiency of the polarizers:

$$CR_{max} = \frac{T_1}{T_2}. \qquad (1.4)$$

The transmittance of practical single polarizers used in LCDs is typically 40–45%, so that $T_1 = 32$–40.5%, and polarizing efficiencies are 99.0–99.9%.

Other types of polarization are circularly and elliptically polarized light. Circularly polarized light can be thought of as a combination of two linearly polarized light waves with equal amplitudes but with a phase difference of $\pi/2$ (90 degrees). In elliptically polarized light, the amplitudes can be different as well.

All liquid crystal displays utilize the optical anisotropy $\Delta n = n_{||} - n_{\perp}$ of the liquid crystal fluid, where $n_{||}$ is the refractive index parallel to the director and $n_{\perp}$ is the refractive index perpendicular to the director. The optical anisotropy causes the polarization components of light to propagate differently, giving rise to birefringence, or double refraction. Birefringence is the key physics phenomenon on which LCD operation is based. The ease with which the director orientation, and therefore the birefringence effects, can be modified in an electric field is responsible for the low-voltage and low-power operation of LCDs.

When natural light enters a slice of LC fluid with uniformly aligned directors (see Fig. 1.9), the light will emerge unchanged only if its propagation direction coincides with the director orientation. When the light is incident under an angle ψ with the director, it can be dissected into two composing linearly polarized waves with different propagation velocities. The first one is called the ordinary wave and it oscillates perpendicular to the plane formed by the director and the propagation direction. The ordinary wave propagates independent of ψ with velocity c/n_0, where c is the speed of light in vacuum and n_0 is the refractive index for the ordinary wave. In the LC fluid, $n_0 = n_\perp$. The other component is called the extraordinary wave and it oscillates in the plane formed by the director and the propagation direction. Its refractive index n_e varies between $n_{||}$ and $n_\perp$, depending on the angle ψ between the director and the propagation direction. Its velocity is c/n_e. For $\psi = 0°$ and $90°$, the refractive index n_e is equal to $n_\perp$ and $n_{||}$, respectively.

The two components will propagate differently and will emerge from the LC slice with a phase change Φ with respect to each other:

$$\Phi = \frac{2\pi(n_e - n_o)d}{\lambda},$$

(1.5)

where d is the thickness of the LC slice and λ is the wavelength of the light.

Since the incident natural light has a random distribution of polarization directions, the exiting light will also be randomly polarized. However, when the light entering the LC layer is polarized perpendicularly to the plane of Fig. 1.9, only the ordinary wave is propagated and the LC cell is said to be operating in the o-mode. When the

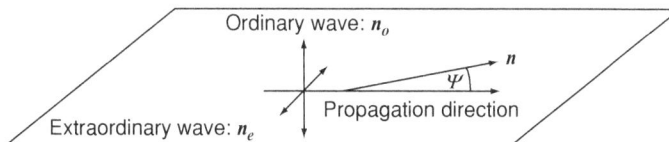

Ordinary wave: n_o

n

ψ

Propagation direction

Extraordinary wave: n_e

Figure 1.9: Decomposition of light into ordinary and extraordinary components, which causes birefringence in optically anisotropic media.

light entering the LC layer is polarized parallel to the plane of Fig. 1.9, only the extra-ordinary wave is propagated and the LC cell is said to be operating in the e-mode.

As an example of how the polarization of light is changed by the LC layer, consider linearly polarized light with wavelength λ perpendicularly entering an LC layer with uniform, homogeneous alignment ($\psi = 90°$) and thickness $d = 1/4\ \lambda/\Delta n$. In this case $n_e = n_{||}$ and the LC slice acts, for all practical purposes, as a quarter-wavelength plate: when the polarization angle is at 45 degrees with the plane of Fig. 1.9, it converts linearly polarized light into circularly polarized light, because one of the two composing waves is retarded by $\Phi = 90°$ (see Eq. 1.5). However, when the orientation of the LC fluid could be changed to homeotropic ($\psi = 0°$) by applying an electric field, the emerging light would remain unchanged and linearly polarized. This example shows the general principle of modifying the orientation of the LC optical axis (the director) by electric fields in order to manipulate the polarization of light. Practical LCDs utilize variations on this basic concept.

1.4   The Twisted Nematic Cell

LCDs are nonemissive (i.e., they need some external lighting source to generate an image). This can be ambient lighting in the case of the reflective display in a calculator, or a backlight in the case of a transmissive display in a notebook computer. The electro-optical behavior of the liquid crystal layer modulates the light from the external light source to form an image or pattern corresponding to the electronic data signal information supplied to the display pixels.

The twisted nematic (TN) mode of operation, which forms the basis of many practical LCDs, was first described by Schadt and Helfrich [2] and Fergason [3] in 1971.

The TN cell basically consists of two glass plates with transparent conductive electrodes patterned on their inner surfaces. A thin LC layer with thickness d = 4–10 μm is sandwiched between the conductive electrodes, as shown in Fig. 1.10. Light enters the LC cell after first passing through a linear polarizer attached to the outside of the display glass. The cell shown in the figure operates in the e-mode, since the polarization direction and the director orientation coincide.

At the surface of the conductive electrodes, alignment layers ensure the proper orientation of the LC molecules so as to obtain a 90-degree twist. In the relaxed state, without applied field, the 90-degree twist of the LC director results in a 90-degree rotation of the polarization direction in the LC cell, in a waveguiding fashion. It can be shown [10] that the light emerges from the cell almost perfectly linearly polarized if the condition $\Delta n.d \gg$ 0.5 λ is satisfied, where λ is the wavelength of the light.

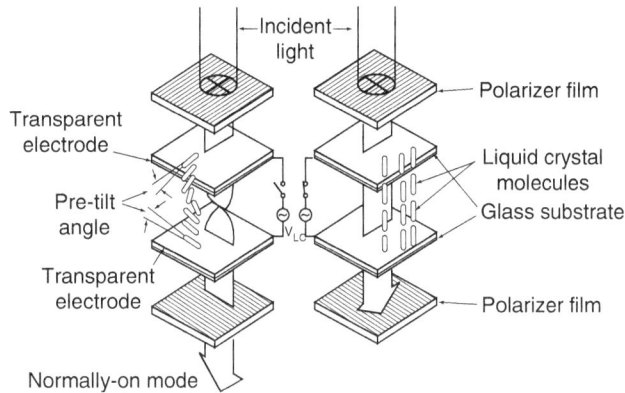

Figure 1.10: Operation of twisted nematic (TN) LC cell.

The second linear polarizer outside the second glass substrate is oriented perpendicular to the first one and will therefore transmit the light. This is called the normally white or normally on mode, since light is transmitted when no voltage is applied to the cell.

The TN effect uses LC fluids with a positive dielectric anisotropy. As a result, the lowest energy state in an electric field occurs when the director is parallel to the electric field direction. When a voltage larger than V_T is applied across the LC, the LC molecules will start deviating from their quiescent state in response to the electric field and the director will tend to line up along the field direction perpendicular to the glass surfaces. This happens initially in the center of the cell farthest away from the transparent electrodes. At the glass surfaces the LC molecules remain parallel to the electrodes. The angle of the director along the z direction is called the director profile. With increasing voltage, more molecules will align parallel to the field. At sufficiently high voltage, the twist is removed and the LC will no longer rotate the polarization direction, so that the exit polarizer now blocks the light. The threshold voltage V_T is defined as the voltage at which the LC molecules start tilting and is different from $V_{th,10}$ in Table 1.1. V_T can be approximated by

$$V_T = \pi \sqrt{\frac{K_{11}}{\varepsilon_0 \Delta \varepsilon}}.$$

(1.6)

It is important to note that in the TN cell, V_T and the entire transmission-voltage curve is, in this first approximation, independent of the cell gap d. In manufacturing, this allows some variation in the cell gap, controlled by spacers, without affecting the gray scale behavior. Such tolerances aid in achieving a high manufacturing yield. The physical reason behind the independence of cell gap can be intuitively grasped: with an increase in cell gap, the electric field causing the LC molecules to line up perpendicular to the glass surfaces is reduced. On the other hand, the influence of the anchoring force to keep the LC director parallel to the glass surfaces is reduced as well. These two forces cancel each other out, as can be mathematically proven, so that the LC director profile is independent of cell gap at a fixed voltage, even though the electric field varies.

When the second polarizer is parallel to the first one, light is blocked in the relaxed (zero voltage) state and transmitted with 4 V applied (the normally black mode). At intermediate voltage levels, a continuous gray scale can be achieved. In Fig. 1.11 the transmission-voltage curves of the TN cell are shown for the normally white and normally black mode.

The LC cell is operated with an AC square wave voltage without a DC component in order to prevent electrochemical degradation of the LC cell structure. DC components in excess of about 50 mV can cause charging of the alignment layer by residual ions in the LC fluid, leading to image retention and other problems in displays. When a periodic waveform $V(t)$ with sufficient frequency is applied to the cell, the transmission curve depends on the root-mean-square voltage V_{rms} of the waveform

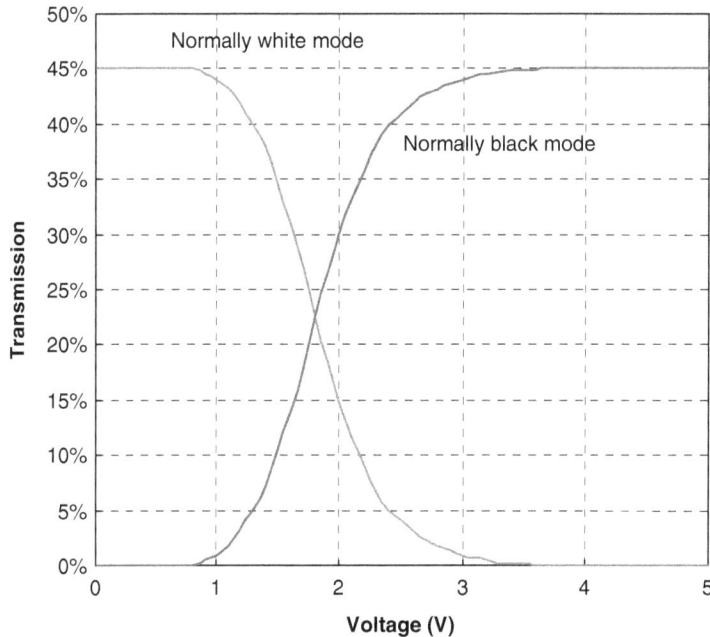

Figure 1.11: Transmittance-voltage curve for TN cell in normally white and normally black mode.

$$V_{rms} = \sqrt{\frac{1}{T_{wf}} \int_0^{T_{wf}} V^2(t)\,dt}, \qquad (1.7)$$

where T_{wf} is the period of the waveform, in displays twice the frame period. In the case of a pure AC square wave voltage with amplitude $+/- V_{amp}$,

$$V_{rms} = V_{amp}. \qquad (1.8)$$

Since the LC molecules are not polar, their orientation does not change when the polarity of the voltage is reversed. In other words, changing rapidly from positive applied voltage to negative applied voltage of the same magnitude at a normal display refresh rate of, for example, 60 Hz has no effect on the director profile. When there are short transient voltages of less than 1 msec applied to the LC layer, the LC response time is usually too slow to react. They normally do not affect the transmission, as long as the RMS voltage does not change significantly.

The TN mode of operation was offered as an example to show how an electric field can change the orientation of the LC molecules and, therefore, the polarization direction and transmission in the cell. There are numerous other LC modes of operation, some of which lead to displays with better viewing angle characteristics. They include the in-plane-switching and vertical-alignment modes, to be described in Chapter 6.

1.5   Limitations of Passive Matrix Addressing

In small LCDs with limited alphanumerical information, such as calculator and electronic watch displays, the transparent ITO electrodes are patterned into segments, as shown in Fig. 1.12. Each segment is individually addressed by the electronics. This approach is practical only for low information content displays.

To obtain LCDs with more picture elements (or pixels), a matrix approach is used. In passive matrix LCDs, the conductive transparent electrodes are patterned as stripes perpendicular to each other on the two opposing substrates (Fig. 1.13). This allows a dot matrix with M rows and N columns (i.e., M×N pixels), to be addressed by M+N external connections. The price to pay for this approach is that the voltage across each pixel is no longer independently controlled. At any time, only one row of pixels is selected and receives data from the column drivers. The data information supplied to the other rows of pixels can, however, modify the average voltage on the pixel.

Dot matrix displays with a one-line-at-a-time driving method, called multiplexing, were developed in the early 1980s. As the number of lines increased, the contrast ratio suffered as a result of the gradual transmission-voltage curve of the 90-degree TN cell. Alt and

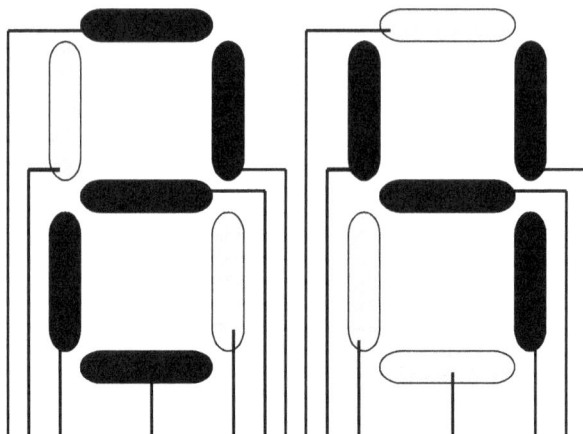

Figure 1.12: Segmented addressing in an alphanumeric display with two digits and seven segments per digit.

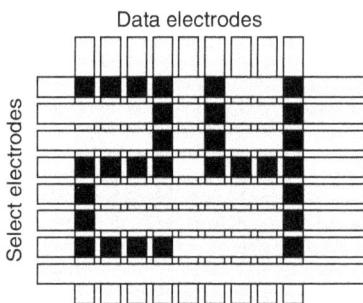

Figure 1.13: Passive matrix addressing.

Pleshko [11] formulated the limits of multiplexing in RMS responding displays. They calculated the optimum voltage ratio between the ON and OFF state as a function of the number of multiplexed lines N:

$$\frac{V_{on}}{V_{off}} = \left(\frac{\sqrt{N}+1}{\sqrt{N}-1}\right)^{1/2}. \tag{1.9}$$

In Fig. 1.14 the Alt–Pleshko equation is graphically represented. In a one-line-at-a-time multiplexing scheme, the voltage margin between the ON and OFF state decreases as the number of rows increases. It results in a reduced contrast ratio unless an LC mode with a

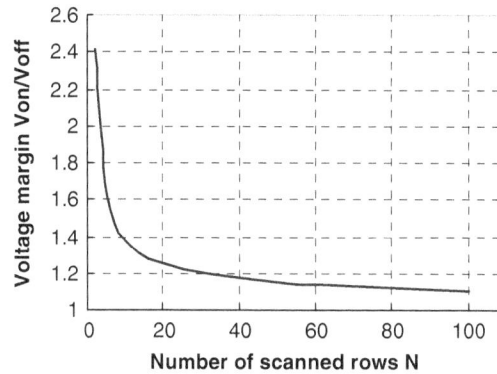

Figure 1.14: Graphical presentation of the Alt-Pleshko equation showing the reduction in contrast ratio with an increase in number of select lines.

very steep transmission-voltage curve is employed. This prompted the development of the super-twisted nematic (STN) cell, which has approximately 270 degrees of twist of the LC layer. It has a much steeper curve, so that a larger contrast ratio T_{on}/T_{off} can be obtained, even when V_{on}/V_{off} is reduced (see Fig. 1.15).

In the late 1980s, STN LCDs were developed. Contrast ratios exceeding 50 have been achieved in 200- to 600-line STN displays. In the higher-resolution displays, such as SVGA types with 600×800 pixels, STN displays are used in a dual scan mode. By dividing the screen in two halves with each half having its own set of data drivers, only 300 lines need to be multiplexed in dual scan STN displays. Multi-line addressing schemes, in which more than one line is selected at a time in STN LCDs, have also been developed; they further improve the contrast ratio and maximum number of addressable rows in passive matrix displays.

Although dual scan STN displays have been applied in early notebook computers, their response time and contrast ratio are relatively poor in comparison with CRTs and active matrix LCDs, precluding high image quality. They are no longer used in laptop displays. The best solution to obtain a high quality, high information content LCD is to add a switch at each pixel and control voltage independently at each pixel to obtain the intended gray level. This emulates a direct-driven pixel and is called active matrix addressing, the subject of the remainder of this book.

The three methods to address LCDs (direct, passive matrix, and active matrix) are listed in Table 1.2 with examples of applications. For all displays with higher information content, a matrix addressing scheme is required to limit the number of interconnects.

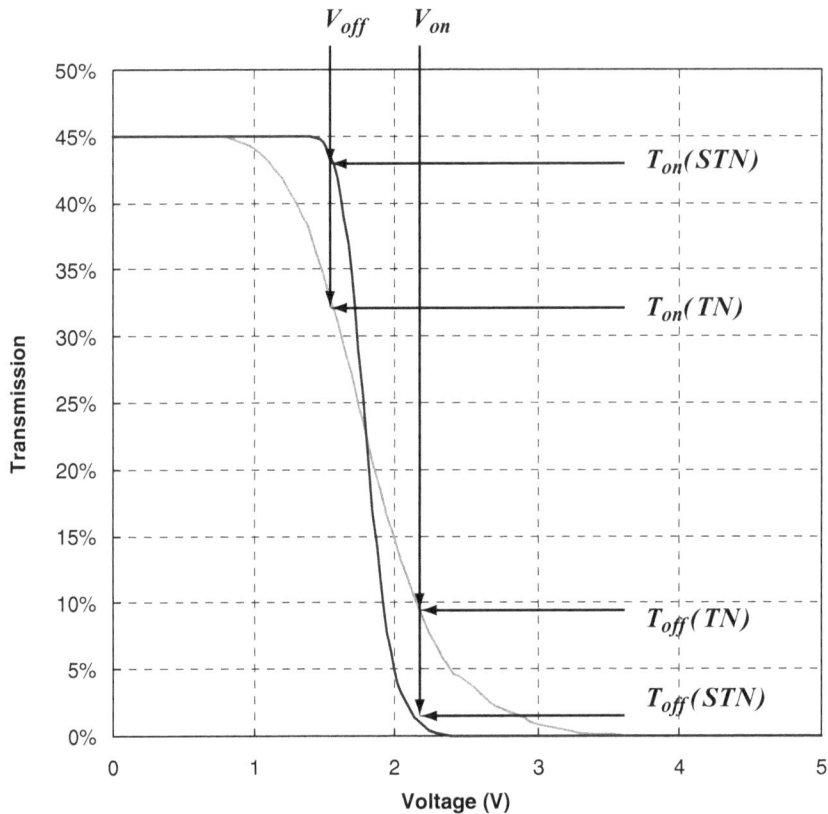

Figure 1.15: Transmittance voltage curve of a TN cell showing the reduction in contrast ratio T_{on}/T_{off} for reduced voltage margin V_{on}/V_{off}. For the STN curve, contrast is much better.

Table 1.2: Addressing methods for LCDs

LCD Type	Addressing Method	Number of Interconnects	Examples
Segmented	Direct drive	One per segment	Watch, calculator
Passive matrix	Multiplexing one row at a time	M + N for M × N matrix	PDAs, cell phones, early notebooks
Active matrix	Switch at each pixel— one row at a time	M + N for M × N matrix	Notebooks, flat panel monitors, LCD TV

References

1. G.H. Heilmeier, L.A. Zanoni, and L.A. Barton, "Dynamic Scattering: A New Electro-Optic Effect in Certain Classes of Nematic Liquid Crystals," *Proc. IEEE*, 56, pp. 1162–1170 (1968).
2. M. Schadt and W. Helfrich, "Voltage-Dependent Optical Activity of a Twisted Nematic Liquid Crystal," *Appl. Phys. Lett.*, 18, pp. 127–128 (1971).
3. J. Fergason, "Display Devices Utilizing Liquid Crystal Light Modulation," U.S. Patent No. 3,731,986.
4. B.J. Lechner, F.J Marlowe, E.O. Nester, and J. Yults, "Liquid Crystal Matrix Displays," *Proc. IEEE*, 59, pp. 1566–1579 (1971).
5. F.J. Marlowe and E.O. Nester, "Alternating Voltage Excitation of Liquid Crystal Display Matrix," U.S. Patent No. 3,654,606.
6. T.P. Brody, J.A. Asars, and G.D. Dixon, "A 6x6 Inch 20 Lines-per-Inch Liquid-Crystal Display Panel," *IEEE Trans. Electron. Devices*, ED-20, pp. 995–1001 (1973).
7. S. Morozumi, "4.25 in. and 1.51 in. B/W and Full-Color LC Video Displays Addressed by Poly-Si TFTs," *SID 1984 Digest*, pp. 316–317 (1984).
8. P.G. LeComber, W.E. Spear, and A. Ghaith, "Amorphous Silicon Field-Effect Device and Possible Application," *Elect. Lett.*, 15, pp. 179–181 (1979).
9. Matsushita a-Si pocket television marketed in 1984.
10. See, for example, L.M. Blinov, *Electro-Optical and Magneto-Optical Properties of Liquid Crystals*. New York: Wiley (1983).
11. P.M. Alt and P. Pleshko, "Scanning Limitations of Liquid Crystal Displays," *IEEE Trans. Electron Devices*, ED-21, pp. 146–155 (1974).

Operating Principles of Active Matrix LCDs

2.1 The Case for Active Matrix

Passive matrix LCDs are relatively easy to manufacture and do not require very large investments. An active matrix LCD, on the other hand, consists of thin film semiconductor circuitry on a glass substrate and shares some of the manufacturing complexity of integrated circuits on semiconductor wafers.

The efficient production of active matrix LCDs requires large capital investments at the same level of wafer fabs for IC manufacturing. The incentive for an LCD producer to make this investment rather than sticking with passive matrix LCDs is simple: image quality, including resolution, maximum size, contrast ratio, viewing angle, gray scale, and video performance is much better in AMLCDs than in passive matrix LCDs. The resolution limitation of the passive matrix was addressed in the previous chapter. Another drawback of passive matrix addressing is that the capacitance of the select and data buslines increases with the square of the diagonal size of the LCD. Since the buslines in passive matrix LCDs consist of transparent conductors with relatively high resistance, the RC delays and power consumption in large passive LCDs become unmanageable.

By adding a semiconductor switch at each pixel, the drawbacks of multiplexing in a passive matrix are eliminated. In the active matrix configuration, the voltage at each pixel and therefore its transmission and gray level can be accurately controlled.

A conventional semiconductor process on crystalline Si wafers could, in principle, be used for AMLCDs. In fact, this has been successful in reflective microdisplays for projection and personal viewers and is commonly referred to as liquid crystal on silicon (LCOS).

For larger displays, however, several factors preclude the use of crystalline silicon. First, it is opaque for visible light and is therefore not compatible with transmissive, backlit displays such as used in notebooks or desktop monitors. Secondly, processed silicon wafers have a limited size (not exceeding 12 in.) and are too expensive. In the semiconductor industry there is a constant push for smaller and more compact circuitry by reducing design rules, which is not needed for AMLCD fabrication.

In an AMLCD most of the viewing area consists of the transparent pixel electrodes. The semiconductor switch and address lines cover only a small fraction of the pixel area. There is no strong incentive to move to advanced submicron design rules since, for large displays with larger pixels, most features actually become larger. In other words, apart from the need for transparent substrates, the use of crystalline wafers would be overkill for the simple circuitry required in most AMLCDs.

Thin film processing has come to the rescue; it allows simple circuitry to be manufactured on large glass substrates with relatively modest design rules and an acceptable, reduced number of process steps.

In terms of operation, the AMLCD is comparable to dynamic random access memory (DRAM). Like the DRAM, the AMLCD consists of arrays of cells in which a voltage is stored. In the case of most LCDs this is not a digital voltage, but an analog voltage to represent different gray levels.

The business model for AMLCD manufacturers also shows some resemblance to that of memory manufacturers. To a large degree, both products have become commodities with constant price pressure and their markets are cyclical in the sense that there are periods of excess supply and excess demand.

2.2   Requirements for Active Matrix Switching Devices

The LC pixel in an LCD can, for all practical purposes, be treated as a low-leakage capacitor. This capacitance needs to be charged from the data voltage at one polarity to the data voltage of opposite polarity during each refresh cycle to obtain an AC voltage without a DC component across the LC pixel. In static images the amplitude of the data voltage across the LC pixels remains constant; only their polarity will change in every frame to prevent degradation of the LC cell. When the displayed information changes (as in video displays with moving images), the amplitude may change as well, so that the gray level can be varied according to the image data.

In Fig. 2.1 the geometry of the LC capacitance C_{LC} for the pixel is shown. It is given by the equation

$$C_{LC} = \frac{\varepsilon_0 \varepsilon_{LC} wl}{d}, \qquad (2.1)$$

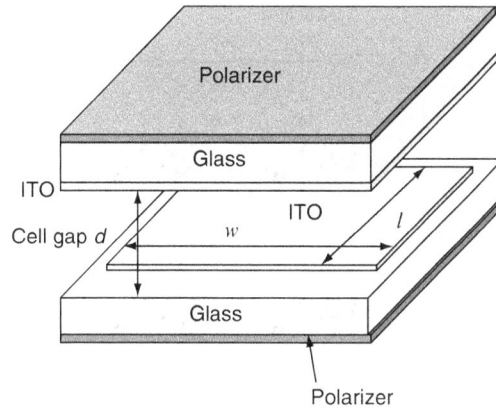

**Figure 2.1: Geometry of the LC pixel
capacitance.**

where ε_0 is the permittivity constant in vacuum, ε_{LC} is the dielectric constant of the LC material, w and l are the side dimensions of the pixel electrode, and d is the LC layer thickness (also called the LC cell gap). Since the LC dielectric constant depends on the orientation of the molecules, the LC pixel capacitance varies significantly with applied voltage.

For example, in a TN cell that uses LC fluid with a positive dielectric anisotropy, the LC director will be perpendicular to the electric field between the pixel electrodes for applied voltages below the threshold. Then, $\varepsilon_\perp$ should be substituted for ε_{LC} in Eq. 2.1.

When 5–10 V AC is applied so that the LC molecules will line up along the electric field direction, $\varepsilon_{\parallel}$ should be substituted in Eq. 2.1. As a result, the capacitance can increase by a factor of two to three. For intermediate applied voltages there is a continuous variation of the effective dielectric constant of the LC layer from $\varepsilon_\perp$ to $\varepsilon_{\parallel}$. LC pixel capacitances in practical LCDs are typically 1 pF or less.

Figure 2.2 shows the circuit of four pixels in a passive matrix LCD. The solid lines represent the circuitry on one substrate, while the dotted lines represent the circuitry on the opposite substrate. In the passive matrix LCD, the LC pixel capacitors are simply formed where the striped transparent electrodes intersect (see also Fig. 1.13 in Chapter 1). The select and data lines are on opposite substrates.

In a TFT-based AMLCD (Fig. 2.3), both select and data lines are on the active matrix substrate, along with the pixel electrode and storage capacitor. The counter-electrode of each LC pixel capacitor, indicated by the dotted line, is a single common transparent

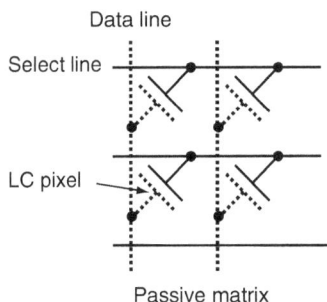

Figure 2.2: Circuit diagram for four pixels in a passive matrix LCD.

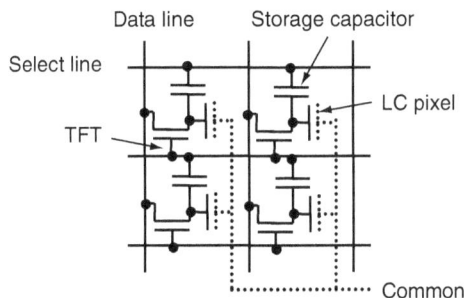

Figure 2.3: Circuit diagram for four pixels in an active matrix LCD.

electrode on the opposite substrate (which usually also has color filters). This common electrode is shared by all pixels in the array.

After the LC pixel capacitance is charged up to the data voltage, the TFT is switched OFF and the pixel electrode on the active array is floating. The pixel capacitor should retain its data voltage for the remainder of the frame time until the next address time. As discussed in Sec. 1.2 in Chapter 1, the LC fluid needs to have a high resistivity to prevent leakage currents. The capacity to retain charge is often expressed in the voltage holding ratio (VHR), which is defined as

$$VHR = \frac{V_{rms}}{V_{peak}},\qquad(2.2)$$

$$V_{rms} = \sqrt{\frac{1}{2T_f}\int_0^{2T_f} V^2(t)\, dt},\qquad(2.3)$$

where V_{rms} is the RMS voltage, V_{peak} is the peak voltage of the LC pixel waveform, and T_f is the frame time, as shown in Fig. 2.4.

In addition to the LC pixel capacitance, most AMLCDs also have an auxiliary storage capacitance connected in parallel to the LC capacitance (as shown in Fig. 2.3). The storage capacitor uses a thin film dielectric with negligible leakage current and its capacitance value is voltage-independent. Therefore, the storage capacitor acts as a buffer to suppress the undesirable voltage dependence and potential leakage current in the LC capacitance. With a storage capacitor at each pixel it is easier to control the RMS voltage on the pixel and therefore its gray level.

In Fig. 2.5 the generic, basic circuit for the switch at each pixel is shown. When one row is selected, all the switches on that row are turned ON and the data voltages are transferred to the pixel electrodes. Ideally, the ON resistance of the switch is zero and the OFF resistance is infinite. In practice, the requirements can be much relaxed. In a display with N rows and a refresh rate of 60 Hz, the frame time T_{frame} to refresh the entire display will be 16.6 msec and the line time T_{line} to address one row will be T_{frame}/N or less. In the case of an XGA display with 768 rows, this translates into a maximum line time of 21 μsec to select one row.

The ON current I_{on} of the switch should be sufficient to fully discharge the LC pixel capacitance and charge it to the opposite polarity voltage V_{on} during the line time. This leads to

$$I_{on} > \frac{2\,(C_{st}+C_{LC})\,V_{on}}{T_{line}}. \tag{2.4}$$

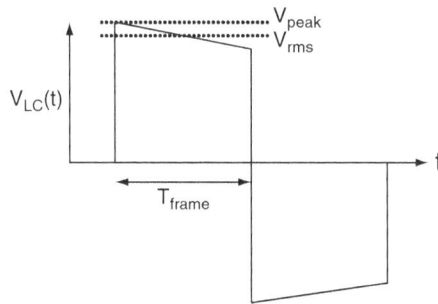

Figure 2.4: Voltage across LC versus time.

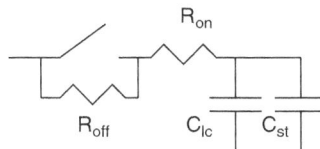

Figure 2.5: Basic switch
configuration in an AMLCD pixel.

The OFF current I_{off} should be low enough to retain the charge on the pixel during the frame time

$$I_{off} < \frac{(C_{st} + C_{LC})\,\Delta V}{T_{frame}},\qquad(2.5)$$

where ΔV is the maximum tolerable voltage loss from the pixel capacitance during the frame time.

When realistic values for a 15-in. XGA display are substituted in Eqs. 2.4 and 2.5, C_{st} = 0.3 pF, C_{LC} = 0.3 pF, and ΔV = 20 mV, we get

$$Ion > 1\,\mu A \; and \; I_{off} < 0.5\,pA.\qquad(2.6)$$

In other words, the ON/OFF current ratio has to exceed six orders of magnitude. The current levels are quite low and can be achieved by transistors using thin film semiconductors such as amorphous silicon (a-Si).

Since the pixels do not draw a current after being charged up during each refresh cycle, the power consumption in the LCD itself is low. Most of the power consumption in transmissive LCDs is usually in the backlight.

Active matrix LCDs can be classified into direct-view displays and light valves (Fig. 2.6). Light valves or microdisplays are used for projection applications and as personal viewers and viewfinders. Microdisplays can be transmissive or reflective. Transmissive types use high-temperature poly-Si TFTs, while reflective types employ crystalline Si circuitry.

Direct-view displays can be further categorized according to the type of switch used at each pixel. The vast majority of AMLCDs on the market use a-Si or poly-Si thin film Transistors (TFTs). Table 2.1 lists some of the pertinent properties of the various

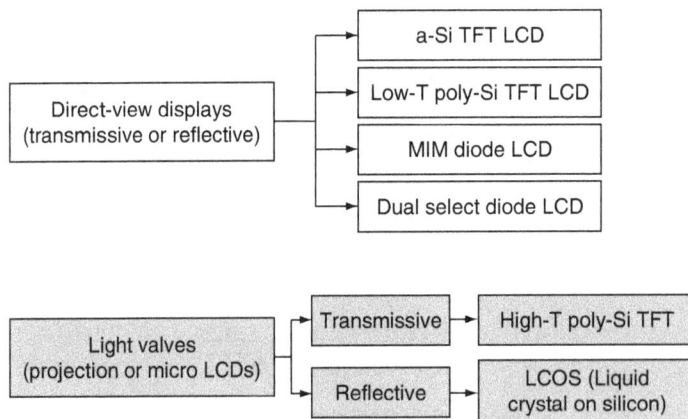

Figure 2.6: Classification of active matrix LCDs.

Table 2.1: Different types of switches for AMLCDs and their main applications

Switching device	Mobility (cm^2/Vsec)*	Highest processing temperature†	Major applications
a-Si TFT	0.3–1	~ 300°C (glass)	Notebooks, flat panel monitors, LCD TVs
High-T poly-Si TFT	100–300	~ 1000°C (quartz)	Projection light valves, viewfinders
Low-T poly-Si TFT	10–200	~ 500°C (glass)	PDAs, notebooks, projection light valves, viewfinders
Crystalline Si MOSFET	400	~ 1100°C (C-Si)	Projection light valves, viewfinders
Thin film diode		< 300°C (glass)	Handheld devices

* Mobility determines the potential to integrate peripheral electronics.
† Highest processing temperature determines substrate choice.

switches, along with display applications in which they are common. The field effect mobility of the TFT (mostly a semiconductor material property) determines the feasibility of designing row and column drivers directly on the glass for a more integrated solution. The highest process temperature determines the type of substrate. For processing below 600°C, low-cost glass substrates can be used. If there is any process step exceeding 1000°C, as in high-temperature poly-Si LCDs, the substrate will be expensive quartz. High-temperature poly-Si TFTs are therefore used only in small displays for projection systems.

Amorphous silicon TFTs have become dominant in large-area displays such as those used in notebook computers, flat panel desktop monitors, and LCD televisions. Low-temperature poly-Si-based LCDs are somewhat less common and are mostly applied in smaller devices, including mobile phones, PDAs, and projection light valves, where their higher processing cost is offset by integration of some of the peripheral electronics on the display glass.

Thin film diodes with ON and OFF currents satisfying Eqs. 2.2 and 2.3 are suitable for LCDs as well. Thus far they have mostly been applied in displays with a limited number of gray levels, such as cell phones.

2.3 The Thin Film Transistor

Thin film transistors (TFTs) are cousins of the transistors used in semiconductor chips. In terms of switching speed and operating voltage, they are inferior to state-of-the-art MOS transistors in crystalline Si. However, they are quite adequate as a simple ON/OFF pixel switch in AMLCDs with a 60-Hz refresh rate. In Fig. 2.7 the top view and cross section of a basic TFT are depicted. The TFT has three terminals: the gate, source, and drain. The gate is separated from the semiconductor film by a gate insulator layer, and the source and drain make contact to the semiconductor. One type of TFT, the N-channel TFT, operates similarly to a field-effect N-channel metal oxide semiconductor (NMOS) transistor in crystalline silicon. It is switched ON by applying a positive voltage on the gate. The insulator acts like a capacitor dielectric, so that in the semiconductor channel an opposing negative charge is induced (see Fig. 2.8). This negative charge creates a conductive channel for electrons flowing from the source to the drain. The magnitude of the current depends on the dimensions of the conductive channel (the channel length L and the channel width W), on the capacitance of the gate insulator per unit area C_g, on the properties of the semiconductor film, and on the applied gate voltage. When a negative voltage is applied to the gate, the channel is depleted of electrons and negligible current flows.

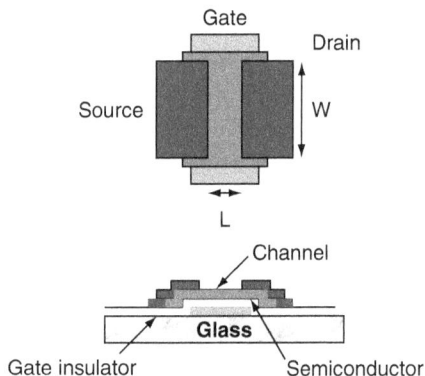

Figure 2.7: Top view and cross section of basic TFT.

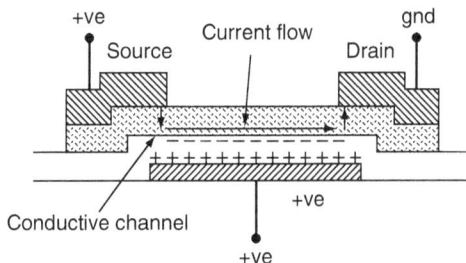

Figure 2.8: Conductive channel formation and current flow in TFT.

The ON current I_{ds} of the TFT can be described, in a first approximation, by the same equations used for traditional NMOS transistors in crystalline Si [1]:

$$I_{ds} = \mu C_g \frac{W}{L} V_{ds}(V_g - V_{th} - 0.5V_{ds}) \text{ for } V_{ds} V_g - V_{th} \tag{2.7}$$

$$I_{ds} = \mu C_g \frac{W}{2L} V_{ds}(V_g - V_{th})^2 \text{ for } V_{ds} > V_g - V_{th}. \tag{2.8}$$

Here μ is the field effect mobility of the thin film semiconductor, a material property, and V_{th} is the TFT threshold voltage, which depends on both the semiconductor and the gate insulator, and on their interface.

Equations (2.7) and (2.8) describe, respectively, the linear and saturation regimes of operation for the TFT, as illustrated in Fig. 2.9.

In the ON state of the TFT, when a data voltage is applied to the source, the drain with the LC load capacitance will charge up to the same voltage, thereby transferring the data

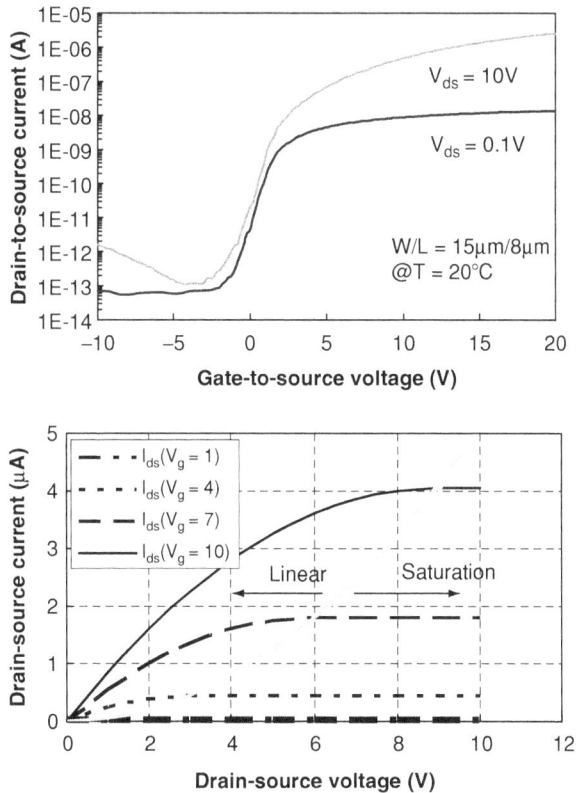

Figure 2.9: Current-voltage characteristics of a-Si TFT.

signal from the data line to the pixel electrode. The TFT is turned OFF when applying a negative voltage to the gate. Since in the N-channel TFT the source and drain contacts block the injection of positive charge carriers (holes), there will be negligible current flow in the OFF state.

Similar equations and considerations apply to P-channel TFTs, in which charge transport is by holes. They are switched ON by a negative gate voltage and switched off by a positive gate voltage.

Many TFTs are of the staggered type, in which the source and the drain have an overlap region with the gate (as shown in Fig. 2.7). The overlap causes a parasitic capacitance that can have an impact on display performance unless the pixel is addressed by an appropriate drive scheme.

The earliest TFTs for LCDs used CdSe semiconductor thin films. Since CdSe is not a standard material in the semiconductor industry, special deposition and etching processes needed to be developed to build CdSe-based TFT LCDs. Although CdSe TFTs showed good performance in AMLCDS, they have never been able to overcome the drawbacks of not being mainstream. They are now relegated to small research efforts.

2.4 Thin Film Silicon Properties

Silicon is a very familiar material in the semiconductor industry and is therefore the preferred choice for use in AMLCDs. Thin films of silicon on glass are, however, inferior in quality to crystalline silicon. This is illustrated in Fig. 2.10, where the atomic structure of c-Si, p-Si and a-Si are schematically compared. In c-Si, the Si atoms are all four-fold coordinated and constitute a perfect crystal lattice with long-range order of the Si atom

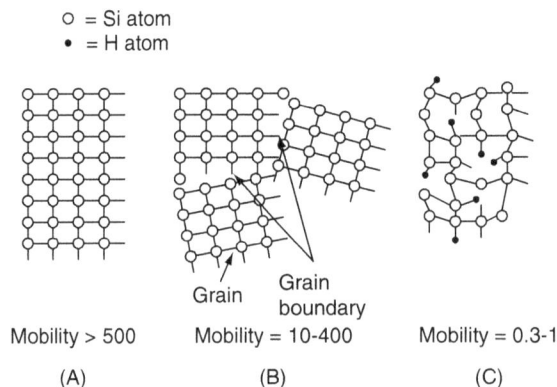

Figure 2.10: Atomic structure of crystalline silicon (A), poly-crystalline silicon (B) and amorphous silicon (C).

positions. As a result, electrons and holes can flow with high velocities v_e and v_p when an electric field E is applied:

$$v_e = \mu_e E \text{ and } v_p = \mu_p E. \tag{2.9}$$

The values of the mobilities μ_e for electrons and μ_p exceed 500 cm^2/Vsec in crystalline silicon.

Low-temperature poly-Si thin films are first deposited as amorphous films, and then crystallized by laser annealing or thermal annealing. The annealing temperature cannot exceed 600°C when low-cost glass substrates are used. Sometimes seed materials such as Ni or preferential crystal growth from certain locations are used. The quality of poly-Si varies greatly depending on deposition and crystallization methods, but invariably it has grain boundaries that reduce the mobility by creating barriers for the flow of electrons and holes (see Fig. 2.10B). Some improvement can be obtained by passivating the grain boundaries with hydrogen in a post-hydrogenation step. The grain size is important in determining TFT characteristics. Preferably, to maximize the mobility, there are no grain boundaries in the channel area of the TFT. A number of techniques have been developed to crystallize the silicon films, including excimer laser annealing (ELA), solid phase crystallization (SPC), sequential lateral solidification (SLS), and metal-induced lateral crystallization (MILC). They result in TFTs with mobilities ranging from 10–400 cm^2/Vsec, the electron mobility often being higher than the hole mobility.

The concept of hydrogenation is important in amorphous silicon as well. The structure of amorphous silicon is more irregular, so that there is no long-range order and no crystal lattice. Nonetheless, since amorphous silicon films are deposited from the decomposition of SiH_4 gas, 5–10% hydrogen is automatically built in during film growth. The hydrogen acts to terminate dangling bonds in the loose a-Si network (Fig. 2.10C) and is essential to improve its electronic properties. The type of a-Si used in TFT LCDs is therefore often referred to as hydrogenated amorphous silicon or a-Si:H. The electron field effect mobility is only 0.3–1 cm^2/Vsec in a-Si, but is adequate for pixel switches in TFT LCDs. The hole mobility is much lower, so that practical p-type TFTs are impossible in a-Si.

Hydrogenated amorphous silicon, as used in LCDs, is a reddish-colored thin film with a thickness between 30 and 200 nm. It can be readily deposited on large glass surfaces by plasma-enhanced chemical vapor deposition. Some of its properties are listed in Table 2.2. It has a band gap of 1.7 eV, (considerably larger than crystalline and poly-crystalline Si [1.1 eV]), a very low dark conductivity, and high photoconductivity. The low dark conductivity makes it relatively easy to obtain a low OFF current in the TFT. The high photoconductivity is an undesirable property of a-Si, since it can cause unwanted photo-leakage currents in the TFT. To prevent photo-leakage, the a-Si channel in TFTs

Table 2.2: Some properties of a-Si films

a-Si film property	Typical value	Unit
Hydrogen content	5–10	%
Dark conductivity intrinsic film	10^{-10}–10^{-9}	$(\Omega cm)^{-1}$
Dark conductivity n$^+$ doped film	10^{-2}–10^{-1}	$(\Omega cm)^{-1}$
Photoconductivity in sunlight	10^{-4}	$(\Omega cm)^{-1}$
Energy gap	1.75	eV
Field effect mobility	0.3–1.0	cm^2/Vsec

is sandwiched between opaque layers to shield it as completely as possible from both ambient light and backlight in the LCD.

Amorphous silicon can be easily doped in the gas phase during deposition, to obtain n-type or p-type films with many orders of magnitude higher conductivity than undoped films. N-type films are used as contact layers to the source and drain of the TFT. The field effect mobility, which is important for the ON current of the TFT (see Eqs. 2.5 and 2.6), is only 0.3–1 cm^2/Vsec and much lower than that of c-Si (500–1000 cm^2/Vsec). Amorphous silicon is therefore used primarily in circuits with relatively low switching speeds (i.e., as the pixel switches in AMLCDs).

2.5 Amorphous Silicon TFTs

Amorphous silicon TFTs have been dominant in notebook and desktop monitor LCDs and in LCD televisions, as a result of their relatively easy processing on very large glass substrates and a limited number of process steps. Amorphous silicon TFTs are N-channel enhancement-type field-effect transistors. As mentioned in the previous section, the electronic properties of a-Si do not allow the fabrication of high-quality p-type TFTs. P-channel a-Si TFTs cannot be made sufficiently conductive to get acceptable ON current for application in displays.

Three different types of a-Si TFTs are shown in Fig. 2.11. The back-channel-etched (BCE) inverted staggered TFT is the most commonly used; it has good performance and a relatively straightforward manufacturing process. Another type is the trilayer (or etch-stopper) TFT. Both the BCE and the trilayer TFT have a bottom gate. The intrinsic a-Si layer on top of the silicon-nitride gate insulator forms the channel.

The source and drain electrodes have an overlap region with the gate and make contact to the a-Si layer via a heavily doped thin n-type a-Si interface layer. The purpose of this a-Si n$^+$ layer is twofold: it acts as a low-resistance ohmic contact for electrons to

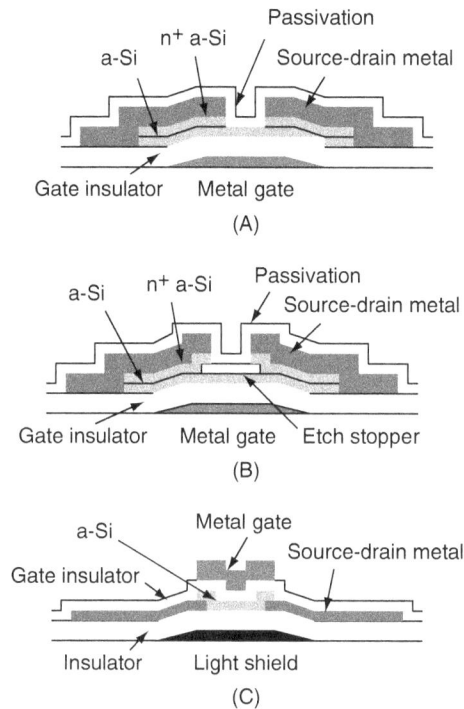

Figure 2.11: Three types of a-Si TFTs: (A) back-channel-etched TFT, (B) etch-stopper or trilayer TFT, and (C) top-gate TFT.

maximize the ON current, and in the OFF state it blocks the injection of holes into the intrinsic i-layer to minimize the leakage current. Since the n^+ layer is conductive, it needs to be removed from the channel area to obtain a low OFF current. This is done by a back-etch after the source and drain electrodes are patterned—hence, the name back-channel-etched TFT. The back-etch process is quite critical and needs to be uniform across the entire manufacturing substrate. Since it is difficult to obtain significantly different etch rates for the a-Si n^+ layer and the undoped intrinsic layer (i-layer), good process control is required to completely remove the n^+ layer from the channel without etching through the i-layer.

In the trilayer or etch-stopper TFT (Fig. 2.11B), the n^+ back-etch is less critical, since there is an extra silicon-nitride etch stopper layer on top of the channel. The etch stopper is deposited and patterned before the n^+ contact layer. Amorphous silicon and silicon-nitride can be selectively etched in dry and wet etch systems. During the back-etch of the n^+ layer there is much better etch selectivity possible against the silicon-nitride etch stopper layer, enhancing the process latitude. The trilayer TFT also has a thinner i-layer, which can suppress photo-leakage currents in the TFT. The major drawback of the trilayer TFT is that it requires an extra deposition and patterning step.

The top-gate a-Si TFT (Fig. 2.11C) is less common. In this configuration the source and drain electrodes are deposited and patterned first, followed by the a-Si layer. The gate insulator (usually silicon-nitride) needs to be deposited on top of the a-Si layer, which usually leads to somewhat lower-performance TFTs in terms of ON current. To suppress photo-leakage currents, a light shield needs to be deposited and patterned prior to the actual TFT process. It is separated from the TFT by an extra insulator layer. The extra processing required for the light shield and the insulator make the top-gate TFT less attractive.

Amorphous silicon is not a very stable material and its structure can be modified by strong illumination or by injection of charge carriers. In addition, the interface between amorphous silicon and the gate insulator (e.g., silicon-nitride) can accumulate charge during operation of the TFT. As a result of this charge trapping, the threshold of the a-Si TFT can shift over time. This effect has been studied in detail, since it can have a bearing on the lifetime of a TFT LCD. The threshold shift versus time can be expressed by the following equation:

$$\Delta V_{th} = \left(V_g - V_{th} \right) * \left(\frac{t}{\tau} \right)^{\beta},$$
(2.10)

where β and τ are constants that can depend on temperature and the quality of the a-Si and a-Si/SiN interface.

The measurement results of threshold shift in non-optimized a-Si TFTs are shown in Fig. 2.12.

Fortunately, the threshold shifts are of opposite polarity during the ON time and the OFF time of the a-Si TFT, so that partial cancellation of the threshold shift occurs. The TFT in an AMLCD functions as an ON/OFF switch to load the pixel data on the pixel capacitor. Therefore, some variation of the TFT threshold across the area is acceptable, as long as the drive scheme can accommodate the variation. It also helps that for a few microseconds after switching ON an a-Si TFT, a transient current occurs that is about 2× higher than the steady-state current. The bottom line is that a-Si TFT threshold shifts have been proven to be small enough to not affect the lifetime of well-designed AMLCDs for up to 50,000 hours.

2.6 Poly-Silicon TFTs

Poly-silicon TFTs (Fig. 2.13) can be subdivided into the high-temperature variety (built on quartz) and the low-temperature type (built on low-cost glass). High-temperature poly-Si TFTs are manufactured with a process that is very similar to semiconductor

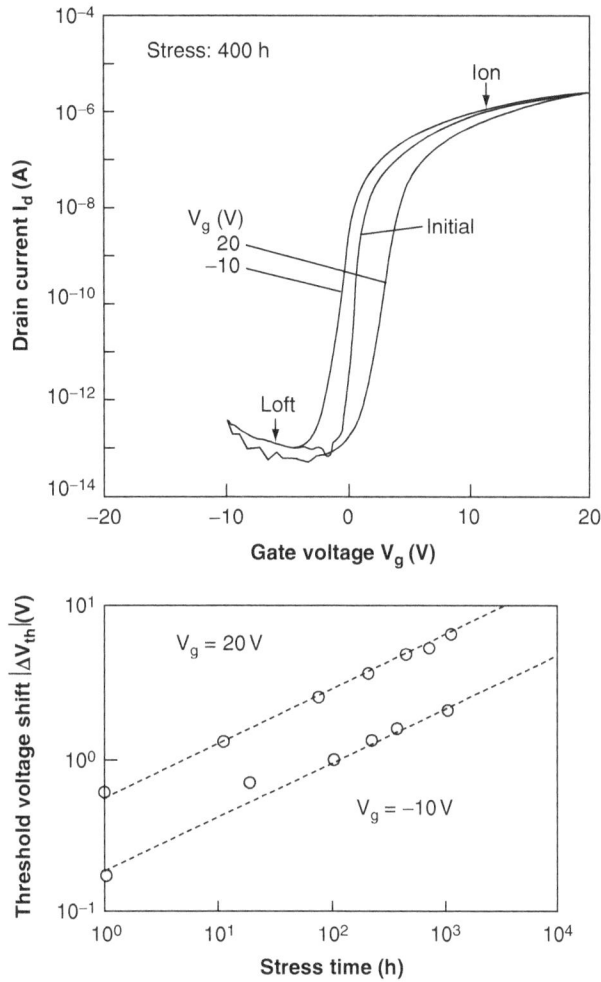

Figure 2.12: Threshold voltage shift in a-Si TFTs after bias stress.

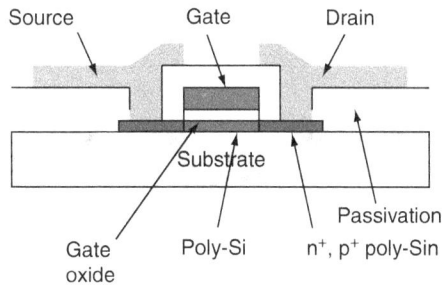

Figure 2.13: Cross section of poly-Si TFT.

processing, with thermally grown oxides, ion implantation, and other steps familiar in semiconductor fabrication. Low-temperature poly-Si TFTs can be produced on large glass substrates with specially developed laser crystallization, ion doping, and gate oxide deposition steps.

With both high- and low-temperature poly-Si, N-channel or P-channel devices can be built so that low-power CMOS circuitry may be integrated along the periphery of the display. The current-voltage characteristics of a-Si and poly-Si TFTs (Fig. 2.14) underscore the difference in mobility of a factor of about 100 and the corresponding difference in ON current. Amorphous silicon TFTs have a low OFF current, sufficiently low to retain charge on the pixel capacitor. Poly-Si TFTs, however, have orders of magnitude higher OFF current, too high for a pixel switch unless special precautions are taken. The high OFF current is caused by high electric fields at the highly doped drain and the imperfect crystalline structure of the poly-Si material. In Fig. 2.15 two methods are shown to reduce the OFF current in poly-Si TFTs. A dual gate structure can be utilized to split the source-drain voltage between two channels and thereby reduce the electric field at the drain. Using an extra process step to create a lightly doped drain (LDD) area has a similar effect. The resulting current-voltage characteristic is shown in Fig. 2.16.

The development and manufacturing of low-temperature poly-Si LCDs has received significant investments because row and column drivers can be integrated on the glass, thereby eliminating the external row and column driver chips. It has been demonstrated that additional circuitry such as D/A converters, DC/DC converters, graphics processors, and even microprocessors can be integrated on the glass as well.

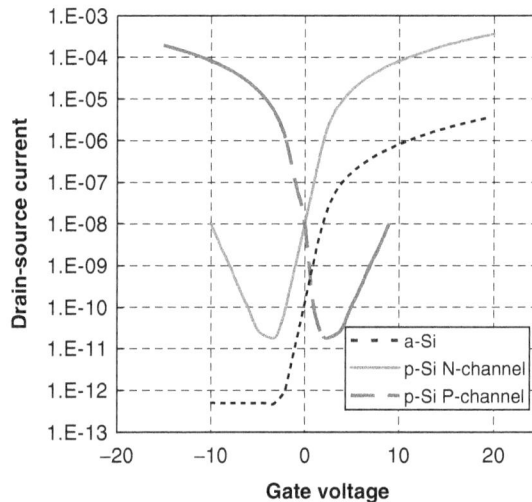

Figure 2.14: Comparison of amorphous silicon and poly-crystalline silicon TFT characteristics.

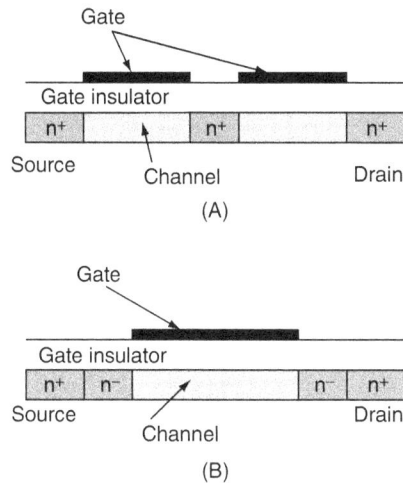

Figure 2.15: Methods to reduce OFF current in poly-Si TFTs: (A) dual gate structure and (B) lightly doped drain structure.

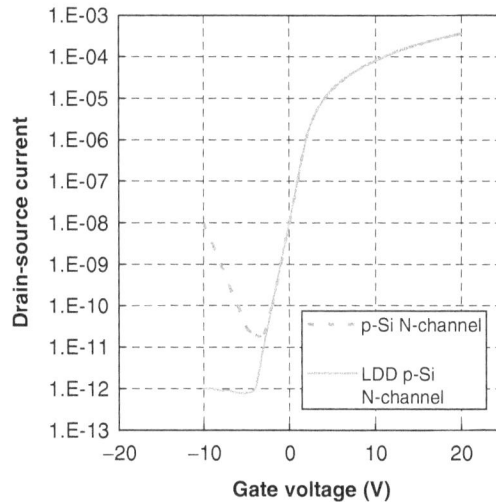

Figure 2.16: Poly-Si TFT characteristics with and without lightly doped drain (LDD) structure.

2.7   Basic Pixel Circuit and Addressing Methods

A TFT pixel array circuit (Fig. 2.17) consists of select buslines, data buslines, a TFT at each pixel, and a storage capacitor in parallel with the LC pixel capacitance. The function of the storage capacitor, as mentioned before, is to act as a buffer and reduce the effects of the

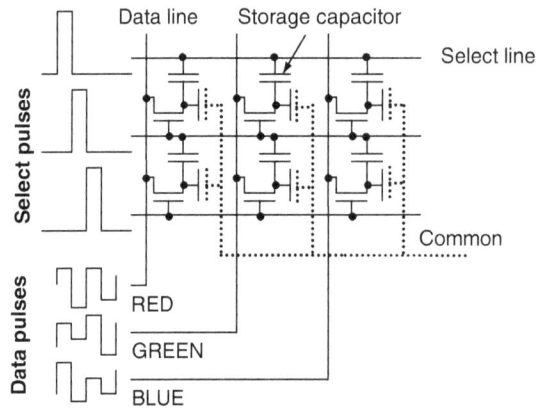

Figure 2.17: Operation of TFT LCD with drive waveforms.

voltage dependence and leakage of the LC capacitance. With a storage capacitor at each pixel, it is easier to obtain uniform gray level performance across the display.

The storage capacitor can be connected to the adjacent select line to maximize aperture ratio in the display. The counter electrode of the LC pixel capacitor is the common ITO electrode on the opposite substrate (the color plate). The display operates by switching on the TFTs one row at a time by a gate pulse with a pulse width of 10–50 μsec, depending on display resolution. The gate pulses are supplied from gate driver circuits, which can either be external or integrated on the glass. While one row is selected, the data signals for that row are applied to the data buslines from peripheral electronic data circuits and transferred by the TFT to the pixel electrodes. When the row is deselected, the TFTs switch OFF and the data voltage is stored for the remainder of the frame time of 16.7 msec (at a 60-Hz refresh rate), while the other rows are scanned. All rows in the display are sequentially selected during the frame time, so that the voltage on each pixel is refreshed once during each frame time. During the next frame time, the polarity of the data signals is reversed, so that all pixels are charged with opposite polarity voltage to obtain the required AC drive without a DC component. For static images, the pixel voltage is simply opposite for odd and even frames, so that the orientation of the LC molecules and the gray levels of the pixel do not change. For moving video images, the amplitude of the pixel voltage can change as well, in addition to its polarity. This causes the LC orientation and therefore the transmittance of the pixel to change over time.

Most AMLCDs are color displays. Color is obtained by using three subpixels per pixel, each having a color filter of one of the three primary colors (red, green, and blue). The color filters are patterned on the top plate opposing the active matrix substrate and separated from the TFT array by the LC layer. The top plate is therefore often referred to as the color plate. A color pixel is obtained by the additive color principle: at close-up, the individual color subpixels may be observed, but from a distance the eye will average the luminance from the three subpixels. When all three colors (R, G, and B) are transmitting, the viewer will perceive the pixel as white; when only R and G are transmitting, yellow will be observed; and combinations of B and G and R and B result in cyan and magenta being displayed.

On the TFT array the three subpixels per pixel are identical, but on the color plate they have different color filters.

A close-up look at the pixel circuit (Fig. 2.18) reveals a number of parasitic circuit elements that can degrade display performance. The parasitics are denoted by the dotted elements. Resistive leakage through the LC or the TFT will both discharge the LC pixel capacitor and affect gray scale uniformity. The parasitic capacitance C_{gd} between the gate and the drain of the TFT causes a voltage division effect with the LC pixel capacitance. This leads to a negative pixel voltage shift ΔV_{pix} of around 1 V on the LC voltage when the gate of an a-Si TFT is switched OFF, as shown in Fig. 2.19.

$$\Delta V_{pix} = \frac{C_{gd}}{C_{gd} + C_{st} + C_{lc}} \Delta V_g, \qquad (2.11)$$

where C_{gd} is the gate-drain capacitance, C_{st} is the storage capacitance, C_{lc} is the LC capacitance, and ΔV_g is the gate voltage change when the row is deselected.

The pixel voltage shift is always negative and could therefore lead to a DC component on the pixel voltage, if the common counter-electrode would be held at the center voltage of the video data signal.

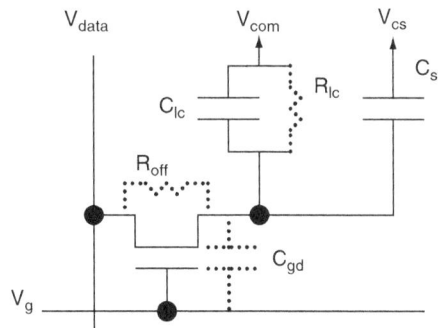

Figure 2.18: TFT LCD pixel circuit with parasitic circuit elements.

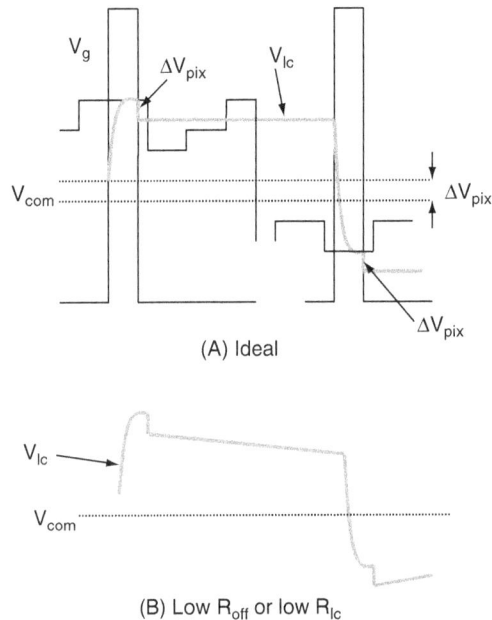

(A) Ideal

(B) Low R_{off} or low R_{lc}

Figure 2.19: Voltage waveforms applied to the select (gate) line and source (data) line with resulting voltage across the LC pixel in ideal case (A) and with leakage through LC or TFT (B).

To eliminate the DC component across the LC, the voltage on the common ITO counter electrode V_{com} is offset by the same voltage to compensate for this pixel voltage shift. As a result of the voltage dependence of the LC capacitance, it turns out to be difficult to completely eliminate the DC component for all gray levels across the entire display area, especially in large displays. Failure to minimize the DC components leads to many undesirable display artifacts, such as image retention, flicker, and non-uniform gray levels. Optimized drive methods are used to reduce the DC component and to prevent it from causing flicker and image retention, particularly in gray scale. Some of these methods will be addressed in Chapter 4.

The storage capacitor in a TFT pixel circuit can be connected to an adjacent gate line or to a specially added common bus (Fig. 2.20). Connection to the gate line maximizes the pixel aperture ratio but tends to increase the load capacitance on the row select lines. This makes it difficult to use this configuration for displays with a diagonal size larger than about 20 in.

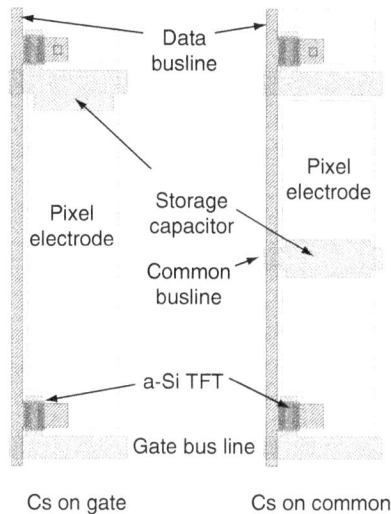

Figure 2.20: Pixel layouts with storage capacitor connected to the gate line (Cs on gate) and with storage capacitor connected to a common bus (Cs on common).

A common storage bus takes up some space in the pixel area and therefore reduces the pixel aperture ratio and the transmittance of the display The RC delay on the gate lines is, however, smaller so that larger displays exceeding 20 in. in diagonal size use this configuration.

2.8   Diode-Based Displays

Instead of a three-terminal TFT, a two-terminal switch or thin film diode (TFD) may be used at each pixel in an AMLCD. There are several ways to implement TFD LCDs. A TFD in this context can be either a rectifying diode or a bi-directional diode, also called a metal-insulator-metal (MIM) diode or a nonlinear resistor. The bi-directional diode has very low current at voltages below about 5 V and starts conducting when a voltage above 10–15 V of either positive or negative polarity is applied.

During the 1980s and early 1990s, several companies worked on rectifying amorphous silicon PIN diodes for AMLCDs. They can be produced with very high rectification (ON/OFF) ratios of eight or more orders of magnitude. At least two rectifying diodes per pixel are needed, one to charge the pixel to positive polarity, the other to negative

43

polarity. In terms of manufacturing, PIN diode LCDs do not have a significant advantage over TFT LCDs and have therefore not been commercially successful.

MIM diode displays have had some commercial success. Their major attraction is a simple, low-cost manufacturing process for the active matrix array with only two or three photo-masks.

In the case of bi-directional MIM diodes, a single diode per pixel can, in principle, be used since it can charge up the pixel to either polarity and is therefore compatible with the AC voltage on the LC capacitor.

In Figs. 2.21 and 2.22, typical current-voltage curves of several bi-directional diodes are shown on linear and semi-logarithmic scales, respectively. The basic circuit and drive scheme of a single TFD LCD are shown in Fig. 2.23. In diode displays, the data lines are ITO stripes on the color plate, as in passive matrix STN LCDs.

Several semi-insulator materials have been used, including Ta_2O_5, Si-rich SiN_x and diamond-like carbon (DLC). The dominant conduction mechanism in bi-directional TFDs is Frenkel–Poole conduction. The Frenkel–Poole effect is the dramatic increase of charge carriers in certain insulators under a high electric field. It is caused by the thermal excitation

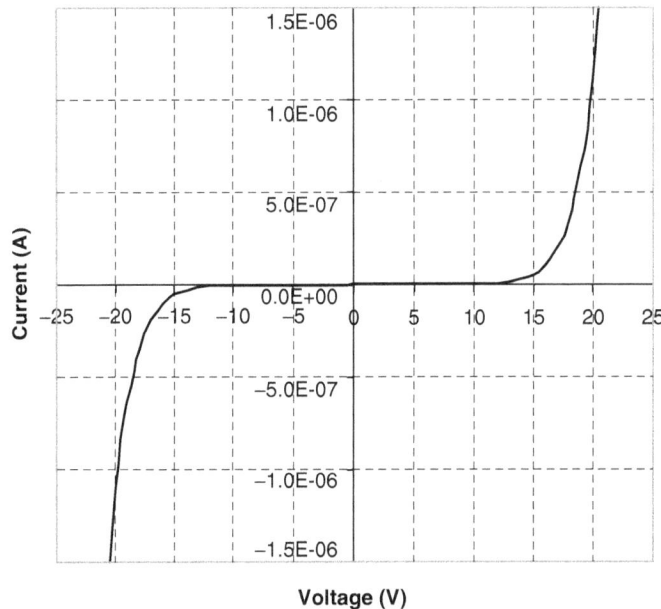

Figure 2.21: Linear current-voltage characteristic of SiN_x diode.

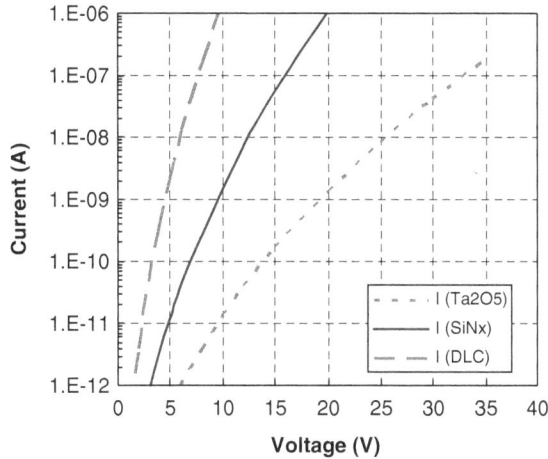

Figure 2.22: Typical semi-logarithmic current-voltage characteristics of Ta2O5, Si-rich SiNx and diamond-like carbon (DLC) diodes (reprinted with permission from the Society for Information Display).

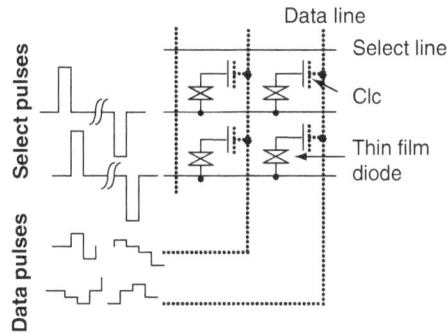

Figure 2.23: Circuit and drive waveforms of a thin film diode LCD.

of charge carriers from traps when the Coulombic barrier for their escape is lowered by the high electric field. This leads to a field-dependent conductivity in the material:

$$\sigma(E) = \sigma_0 \exp(\alpha \sqrt{E}), \qquad (2.12)$$

where σ_0 is the zero field conductivity and α the nonlinearity coefficient:

$$\sigma_0 = q\mu n \exp\left(-\frac{\Phi_b}{kT}\right), \qquad (2.13)$$

45

$$\alpha = \frac{1}{kT}\sqrt{\frac{q^3}{\pi\varepsilon_0\varepsilon_r}}. \tag{2.14}$$

Here T is the absolute temperature, k is Boltzmann's constant, q is the electronic charge, ε_0 is the permittivity in vacuum, ε_r is the dielectric constant, Φ is the trap depth for the charge carrier, μ is the mobility of the charge carriers, and n is the carrier density. The dielectric constants for Ta_2O_5, Si-rich SiN_x and diamond-like carbon (DLC) are, respectively, 24, 9, and 4.

The best diodes are obtained with a semi-insulator with a low dielectric constant ε_r, as follows from Eq. 2.14. A low dielectric constant also minimizes the diode capacitance, so that cross-talk is reduced. From this perspective, DLC appears to be a good candidate for further development. Ta_2O_5 diodes have been widely used in small mobile displays and are obtained by anodizing Ta. Unfortunately, Ta_2O_5 has a high dielectric constant and relatively poor current-voltage characteristics, which precludes large, high-resolution displays. Si-rich SiN_x has the advantage that it can be processed in standard PECVD equipment familiar in every a-Si TFT LCD line.

The diode current in the nonlinear resistor can be derived from Eq. 2.12:

$$I = S\kappa V \exp\left(\beta\sqrt{|V|}\right) \tag{2.15}$$

$$\beta = \frac{\alpha}{\sqrt{d}}, \tag{2.16}$$

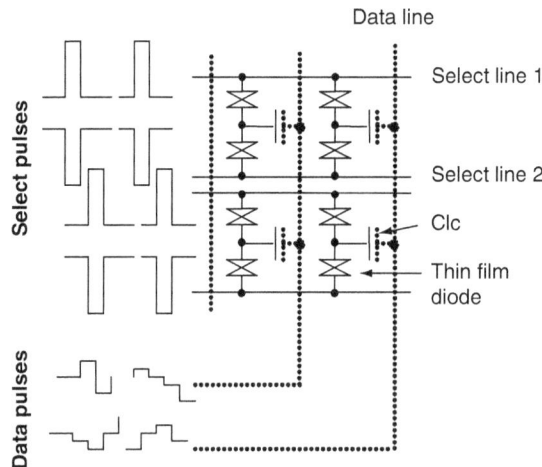

Figure 2.24: Circuit and drive waveforms of a dual select diode AMLCD.

where S is the diode area, V is the applied voltage, d is the thickness of the insulator in the TFD and κ and β are material parameters which depend on temperature and on the dielectric constant of the insulator film.

The TFD and the LC capacitor are in series in Fig. 2.23 and this makes the LC voltage extremely susceptible to small variations in the diode characteristics over time and across the display. Unfortunately, it is difficult to make TFDs very uniform over large areas. In addition, the insulator typically used in MIM diodes (anodized Ta_2O_5) leads to diodes with marginal ratios between ON and OFF current.

The series connection of the diode and LC capacitor also causes a strong dependence of the LC voltage on signal propagation delays on the buslines, another factor limiting gray-scale performance, especially for large-diode LCDs.

Single-MIM-diode displays have therefore been only successful in small applications such as cell phones with a limited number of gray levels.

An approach that eliminates the drawbacks of the single-MIM-diode pixel circuit is the dual select diode (DSD) AMLCD, which has a pixel circuit as shown in Fig. 2.24. By applying opposite polarity pulses to the two select lines, the pixel electrode voltage is accurately reset during each select time.

Unlike in the single-TFD circuit, variations in the TFD across the display area and over time are basically cancelled out in the differential DSD pixel circuit. The propagation delays on select and data addressing pulses are also cancelled out to a large degree in DSD LCDs. This leads to much more uniform gray levels and the potential to scale up to large-area displays with diagonal size exceeding 30 in. This technology is under development.

2.9 Plasma-Addressed LCDs

During the 1990s a totally different technology for AMLCDs was developed, based on a plasma-addressing technique rather than on transistor or diode switches.

Plasma-addressed liquid crystal displays (PALC displays) were the first type of AMLCDs with a diagonal size over 40 in. They employ a hybrid technology with plasma channels used to address one row at a time. The plasma channels are separated from the LC layer by a microsheet of glass with a thickness of only 50 μm. The pixel electrodes are reset to a predetermined voltage during each select time. PALC displays have ITO data lines patterned on the color substrate. The color plate process for PALC displays is similar to that in STN passive matrix LCDs and diode LCDs, which also have ITO data lines on the color plate.

In PALC displays the reset operation of each pixel is performed by a plasma channel between a cathode and an anode busline. The plasma channels are used only for addressing. Display luminance is controlled independently by a backlight. PALC displays require high voltage select drivers.

The combination of plasma technology with LCD technology for PALC displays considerably increases process complexity. Issues with PALC displays include varying decay times of the plasma, bending of the microsheet between the plasma and the LC layer, and process integration problems. This technology has therefore not matured to mass production.

References

1. S.M. Sze, *Physics of Semiconductor Devices*. New York: Wiley (1969).

Manufacturing of AMLCDs

In this chapter the steps in the basic manufacturing process of AMLCD panels will be described. Panel manufacturing is usually subdivided in TFT array manufacturing, LC assembly, and module assembly. TFT array production shows some similarities with semiconductor manufacturing. Many of the methods, types of equipment, and expertise acquired in the semiconductor industry have therefore been successfully applied to AMLCD array processing. They include yield optimization techniques and clean room and facility operation, and are applied to the color filter plate manufacturing as well. The front end of AMLCD manufacturing is highly automated with expensive equipment and is therefore very capital-intensive. Back-end processing (module assembly and assembly of the final product) is more labor-intensive and has often been moved to low-labor-cost regions in Asia.

3.1 Basic Structure of AMLCDs

In Fig. 3.1 a cutout view of a basic TFT LCD panel is shown. It consists of a TFT array glass plate and a color filter array plate. The two are separated by 4–5 μm and form the sandwich for the liquid crystal layer. At the periphery of the display viewing area (Fig. 3.2), a glue seal bonds the two glass plates together and also prevents moisture and contamination from entering the LC fluid.

The row and column drivers are attached to ledges at the edges of the TFT array glass plate to supply the address signals to the LC pixels. In a-Si TFT LCDs they are packaged on a flexible tape in a so-called "tape carrier package" (TCP) and attached to the glass by tape automated bonding (TAB). Alternatively, the row and column drivers may be attached upside down directly to the glass with "chip on glass" (COG) technology. The row and column drivers receive their signals from a printed circuit board with controller circuitry, including a timing controller. The panel is completed by adding a number of optical films to enhance the viewing properties and uniformity, a backlight assembly, and an enclosure.

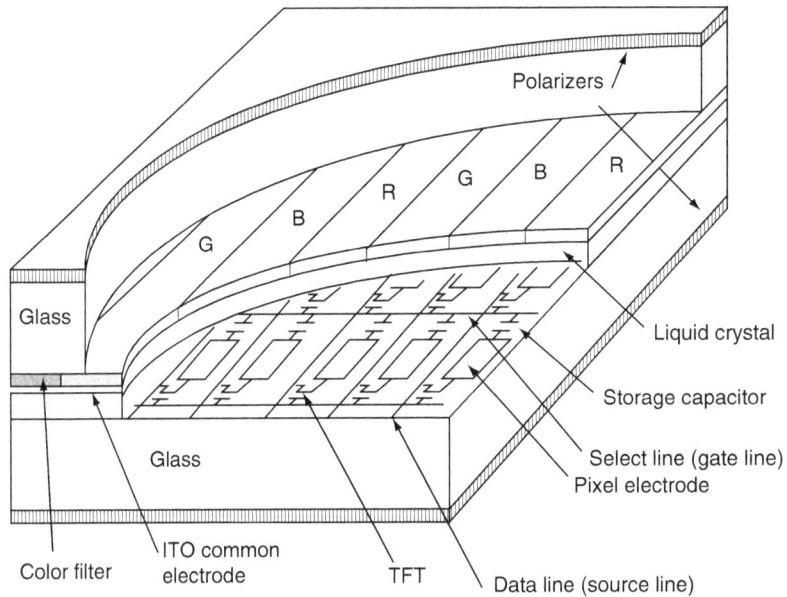

Figure 3.1: Cutout view of a TFT LCD.

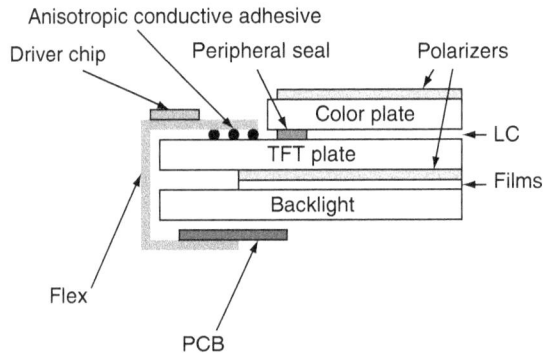

Figure 3.2: Cross section of an edge of a TFT LCD.

3.2 Thin Film Processing

The pixel array circuitry is patterned on one of the two glass substrates between which the LC layer is sandwiched. Since the thickness of the liquid crystal layer in an AMLCD is only 4–5 μm, the active matrix consists of thin film circuitry with a total thickness usually less than about 1 μm. Fortunately, thin film processing of conductors and insulators is well established in the semiconductor industry. However, until the 1980s patternable thin film semiconductors on glass were much less common. They are needed because opaque crystalline wafers cannot be used for transmissive displays with a backlight. Moreover, semiconductor wafers are too expensive and limited in size to not more than 12 in.

Amorphous and polycrystalline silicon thin films turn out to be quite suitable for AMLCDs and are easily patternable using etch chemistry derived from silicon processing in the semiconductor industry. Manufacturing of a-Si TFT arrays for LCDs requires specialized equipment for vacuum deposition, etching, and patterning of thin film layers.

RF sputter deposition in Ar gas with metal targets (Fig. 3.3) is used for the select and data busline metals, which also function as the gate and source/drain electrodes, respectively. The indium-tin oxide (ITO) transparent conductor for the pixel electrodes is deposited with reactive sputtering in Ar and O_2.

The key ingredients for the TFT itself are the gate insulator (Si_3N_4) layer and the semiconductor (a-Si) layer, which are both deposited in vacuum (Fig. 3.4) by plasma-enhanced chemical vapor deposition (PECVD). The glass substrates are heated to 300–350°C.

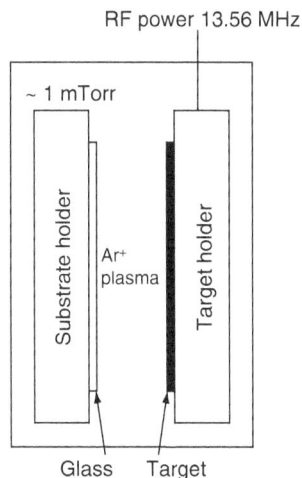

RF power 13.56 MHz

~ 1 mTorr

Substrate holder

Target holder

Ar+ plasma

Glass   Target

Figure 3.3: Schematic view of sputtering system for metal and ITO deposition.

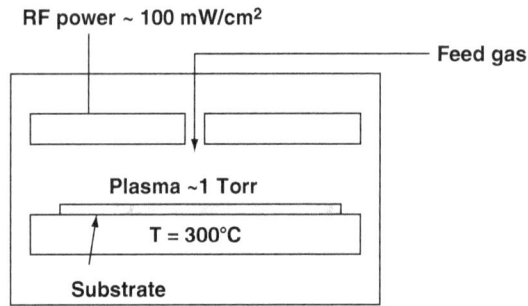

RF power ~ 100 mW/cm²

Feed gas

Plasma ~1 Torr

T = 300°C

Substrate

Figure 3.4: Schematic view of plasma enhanced chemical vapor deposition (PECVD) equipment.

The thin films are grown on the heated substrate by the decomposition of a feed gas in RF plasma, amorphous silicon from SiH_4 and H_2 gas at a low pressure below 1 Torr (1 mm Hg). For the doped n^+ layer, which acts as a low-resistance ohmic contact between the source/drain and the a-Si layer, a small amount of PH_3 gas is added to the mixture to obtain phosphorous doping. The Si_3N_4 gate insulator is grown in SiH_4, NH_3, and N_2. In Table 3.1 the various PECVD layers are listed with their feed gases and function.

The equipment to deposit the PECVD layers is a cluster tool (Fig. 3.5) with a central robot in vacuum surrounded by heating, processing, and load/unload chambers. A similar type of vacuum equipment is also used for dry etching the PECVD layers and some of the metals. The other metals and ITO are patterned by wet etching.

The various layers are patterned with photolithography to obtain a TFT array and a grid of data lines and select lines. Photolithography is a well-known process in the semiconductor industry and the method has been adapted for the patterning of layers in TFT arrays. Unlike in the IC industry, the challenge in flat panel lithography is not so

Table 3.1: PECVD materials used in AMLCDs

Feed gas	Material	Function
SiH_4, H_2	a-Si	Semiconductor
SiH_4, N_2, NH_3	Si_3N_4	Gate insulator, passivation
SiH_4, N_2O	SiO_2	Gate insulator, passivation
SiH_4, PH_3, H_2	n^+ a-Si	Contact layer at source and drain

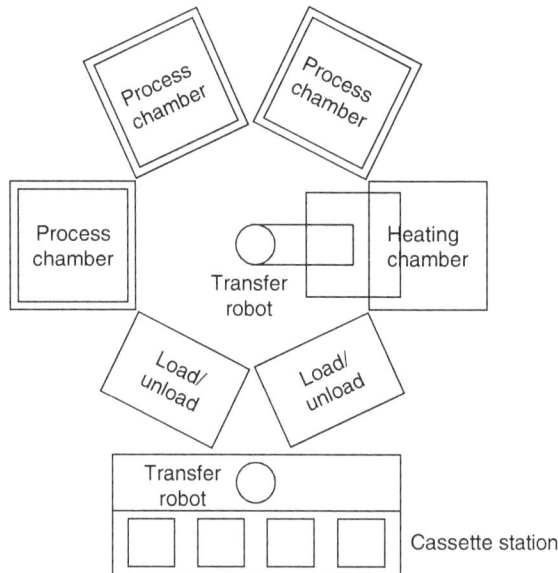

Figure 3.5: Cluster tool configuration for PECVD of a-Si and Si$_3$N$_4$.

much the ultra-fine patterning of the layers, but more the capability to pattern on large glass substrates with medium resolution and high throughput.

The generic sequence of a patterning step is as follows (Fig. 3.6).

1. Deposition of thin film layer
2. Substrate cleaning with deionized water
3. Photoresist coating and baking
4. Exposure of the resist through a mask
5. Photoresist developing and baking
6. Wet or dry etching of the layer to be patterned
7. Photoresist stripping

Cleaning is done by spraying in deionized water and spinning to dry. For large substrates, spraying is followed by an airknife to dry.

Until recently, the photoresist layer of about 1 μm was exclusively applied by spin coating.

This procedure consists of dispensing a puddle of resist on the center of the substrate followed by spinning the substrate (Fig. 3.7A). The substrate holder is rotated at high speed, which works well for smaller substrates (less than about 1 m × 1 m). For larger, heavier glass substrates, high-speed spinning poses major problems. In addition, in the spin-coating process much of the photoresist is lost from spinning off the substrate. Since the material utilization rate of spin coating is so poor (less than 30%), newer techniques

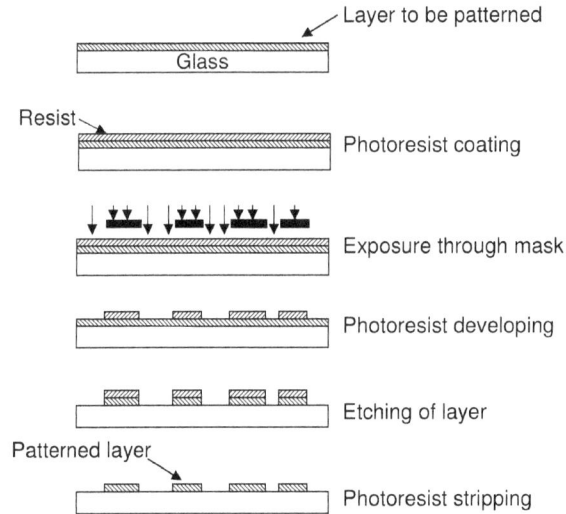

Figure 3.6: Photolithography sequence.

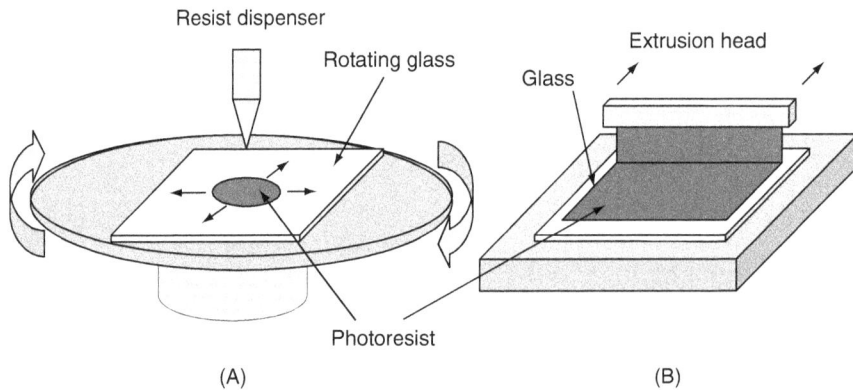

Figure 3.7: Principles of spin coating (A) and slit (extrusion) coating (B).

for very large substrates use a combination of extrusion or slit coating through a narrow slit, followed by spin coating or just extrusion coating only (Fig. 3.7B).

Various types of exposure equipment are used to selectively expose the resist to blue light or UV radiation generated by a Hg (mercury) arc lamp, including flat panel projection steppers (Fig. 3.8), mirror scanning projection aligners (Fig. 3.9), and proximity aligners (Fig. 3.10).

In steppers and mirror projection aligners, the mask (containing a master copy of the pattern) and the substrate are well separated from each other and an optics system projects the pattern from the mask on the substrate. The mask is a quartz plate with a zero-defect Cr pattern based on the layout design of the particular layer.

In the stepper, the optics system includes a projection lens optimized for one wavelength of the mercury arc lamp spectrum (usually the g-line at 436 nm). An accurately controlled XY stage moves the substrate into position for exposure. Unlike in semiconductor wafer exposure systems, there is normally no reduction of the image from mask to substrate. In some equipment, magnification of 1.25x to 2x is actually used.

In the mirror projection aligner (Fig. 3.9), there is no projection lens, but the image is transferred from the mask to the substrate by a set of flat and concave mirrors. The exposure light is formed into a slit shape and the mask and substrate move together in

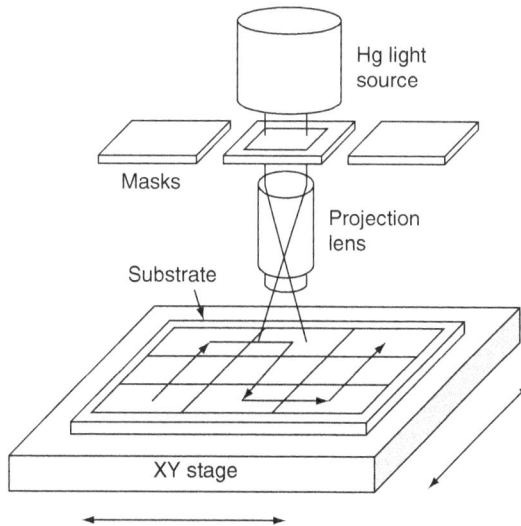

Figure 3.8: Flat panel stepper operating principle.

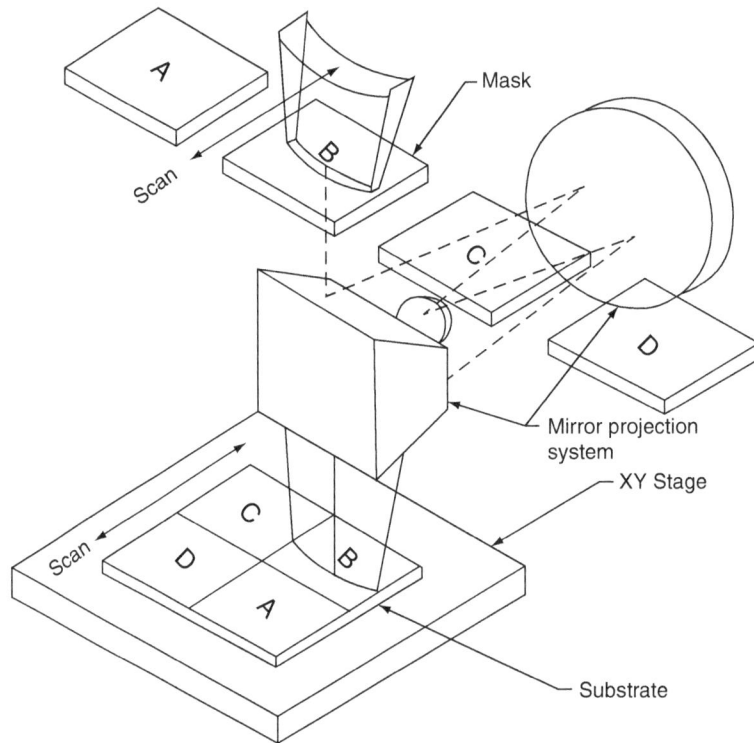

Figure 3.9: Mirror projection aligner operating principle.

such a way that the slit scans across the mask and the substrate. As a result, there is no reduction or magnification. Since there are no refractive lenses in the projection system, there is no concern about chromatic aberration in mirror projection aligners. The mirror projection aligner can therefore use all three major lines of the mercury lamp spectrum (the g-, h-, and i-lines at 436, 405, and 365 nm, respectively).

Proximity aligners (Fig. 3.10) have a simpler structure with a uniform radiation source illuminating the mask and substrate, which are separated by a proximity gap of only 10–30 μm. The Cr pattern on the mask is on the bottom side of the mask close to the substrate. As a result, the pattern on the substrate will be the mirror image of the pattern on the mask. Maintaining a constant gap between the substrate and mask is one of the main challenges for proximity aligners. The other is the prevention of larger particles on both substrates and masks, which could adhere to the mask during the exposure cycle. These larger particles can impact yield, require cleaning, or make the mask useless.

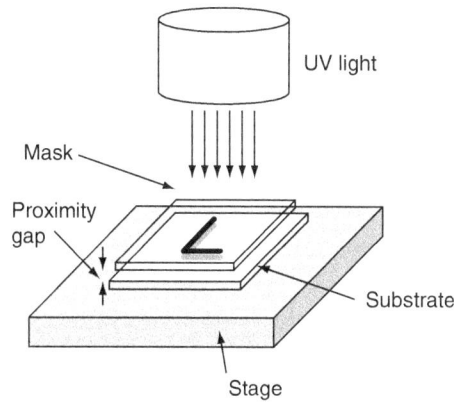

Figure 3.10: Proximity aligner operating principle.

The resolution and layer-to-layer overlay accuracy (4 μm and 1 μm, respectively) of steppers and mirror projection aligners are sufficient for patterning the TFT array. It is interesting to note that for a-Si TFT array patterning, the design rules do not change much with display size or resolution for direct-view displays. This is in stark contrast to the perpetual race for smaller design rules in semiconductor chips (now less than 0.1 μm in state-of-the-art devices). This is related to the fact that direct-view displays need to have certain dimensions and actually have a tendency to grow in size. In particular, for LCD television the pixel size becomes larger for larger displays. However, design rules are not significantly relaxed for patterning large-size TFT arrays for direct-view displays. They all need about 4-μm minimum feature sizes and 1-μm overlay accuracy to maintain uniform gate-drain capacitance across the display area (see Eq. 2.11).

The flat panel display industry uses positive photoresist for the TFT array (i.e., the exposed part of the resist is removed during developing). The underlying material is then etched to replicate the remaining photoresist pattern and the pattern on the mask. The minimum feature sizes and layer-to-layer overlay accuracy for patterning the TFT array and the color plate are listed in Table 3.2.

The key parameter determining productivity of the expensive exposure equipment is throughput. For Generation 5 and higher plants, step-and-scan projection aligners using large quartz masks are dominant to enhance throughput (as shown in the example of Fig. 3.9).

For lower-volume production and older-generation plants, the very high cost of the large quartz masks for step-and-scan projection aligners can be a stumbling block. In that case,

Table 3.2: Design rules for TFT array and color filter array

Process	Resolution	Registration/ Overlay
TFT array	4 μm	1 μm
Black matrix of CF array	4 μm	1 μm
CF array	~ 8 μm	2 μm
CF to TFT array		< 3 μm

projection steppers are preferred with a 1x projection ratio and small masks of size 7 in. × 7 in. or less. The mask in flat panel projection steppers can be smaller than the size of a notebook panel so that the patterns need to be stitched together by a sequence of exposure shots, as shown in an example in Fig. 3.11.

Stitching can cause visible artifacts in the AMLCD image unless the variation in registration accuracy of the layers across stitching boundaries is within certain limits. In particular, the gate-drain overlay variations and pixel ITO pattern registration can lead to small brightness variations at the stitching boundaries. These artifacts are referred to as stepper patterns or "shot mura" (after the Japanese word *mura*, which indicates non-uniformity). The word "mura" is also used for other types of non-uniformities and cosmetic defects in the display image resulting from non-optimized processing.

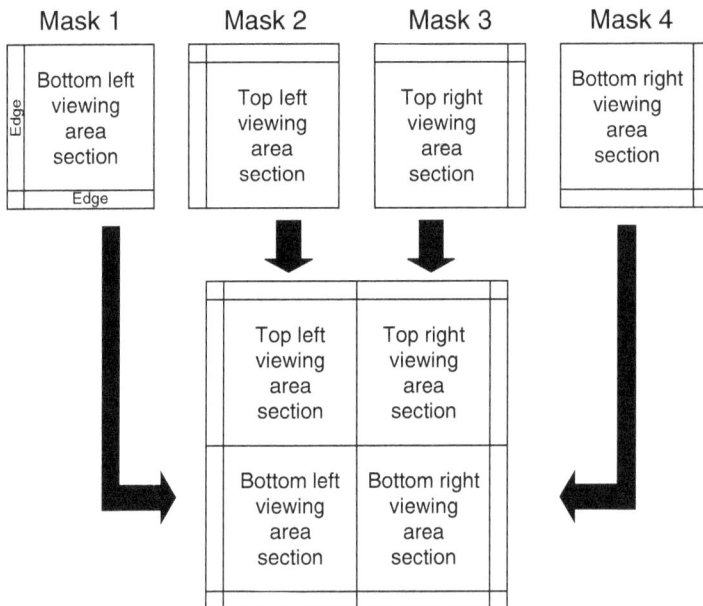

Figure 3.11: Example of stitching of patterns in a flat panel stepper.

The developing of the photoresist is done by spraying or by immersion in the developer. The etching step removes the conductor, insulator, or semiconductor thin film selectively where the photoresist has been developed away. Either wet etching or dry etching is possible.

The ITO layer and some of the metals are wet etched in acid solutions adapted to the type of material. Wet etching processes are isotropic (i.e., the etch rate is the same in all directions). Therefore, overetching can lead to severe undercutting of the layer to be patterned under the photoresist (Fig. 3.12A).

The semiconductor and insulator layers (and sometimes the metal films) are dry etched in reactive ion etching (RIE) systems using a plasma of etching gases such as SF_6 or CF_4. RIE is more anisotropic than wet etching (Fig. 3.12B) and can therefore better maintain the dimensions of the photoresist pattern.

By adding O_2 to the gas mixture, the photoresist can be partially eroded by dry etching to obtain a tapered edge profile in the patterned film (Fig. 3.13). This facilitates step coverage of later layers over the pattern. For example, the data line metal needs to step over the gate line metal at crossover points in the matrix so that the gate line metal preferably has a tapered edge. Although dry etching needs more expensive, vacuum-based equipment than wet etching, there is a trend to move to dry etching for most layers to maximize yield and throughput.

An important parameter is the etch selectivity to buried patterns under the layer to be patterned. Normally, the layers patterned earlier in the process should not be attacked by the etchant for subsequent patterning steps. In this respect, wet etching is often better than dry etching, since it is easier to find a wet etching solution that does not attack underlying layers than a selective etching gas.

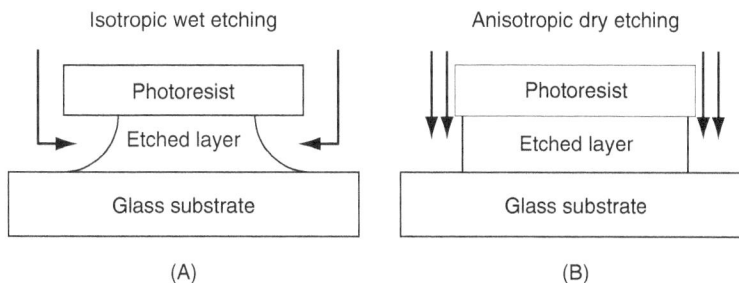

Figure 3.12: Edge profile of patterned layer after wet etch (A) and dry etch (B).

O$_2$ containing etching gas

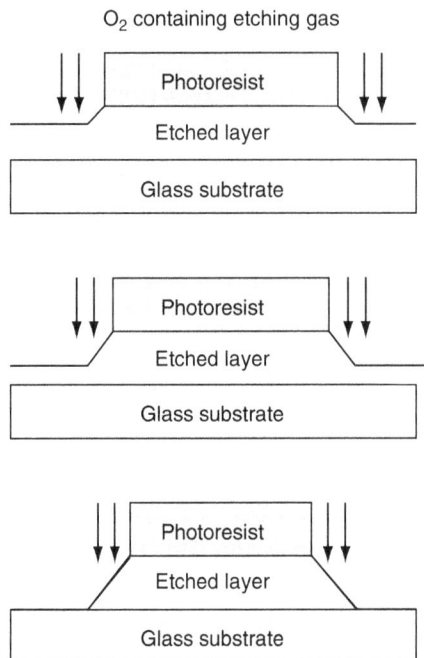

Figure 3.13: Tapered edge profile obtained during dry etching with photoresist erosion.

After etching the thin film, the photoresist is removed by immersion in a stripping solution or by dry stripping in oxygen containing plasmas. The cleaning steps in the photolithography process are performed by spraying deionized water on the substrate, followed by drying through spinning or with an airknife.

The color filters on the color plate do not require high-resolution and registration accuracy (see Table 3.2) and are therefore patterned with lower-cost proximity aligners. The red, green, and blue filters are deposited by extrusion coating followed by spinning, or by extrusion coating only to better utilize the color filter material. Today's color filter materials are mostly photo-imageable (i.e., they have a photo-initiator added and can be developed after exposure without the need for additional photoresist coating and stripping steps). They are mostly negative-resist-type materials with color pigment added. Because the exposed part of the color filter layer will remain after developing, the master pattern on the mask must be a negative of the color filter pattern.

3.3 Thin Film Properties

The thin layers in the active matrix array are either conductor, insulator (dielectric), or semiconductor films. Polymer films are sometimes used as well. The films must satisfy a number of electrical, optical, and mechanical requirements to be useful in LCDs. An additional requirement is the ability to efficiently deposit and pattern the layers at acceptable cost on glass substrates with excellent yield.

The conductor films may be subdivided in metal layers that are patterned as the TFT electrodes and addressing buslines, and in transparent conductors such as indium-tin oxide (ITO) that are delineated as the pixel electrodes.

For the metal conductors, refractory metals (e.g., Cr, Ta, and Mo) were originally used. Although adequate for small displays, their resistivity ρ is too high for large displays exceeding about 15 in. in size. Table 3.3 lists the resistivities of some conductor films used in AMLCDs, along with their functions.

In thin films the resistivity is generally higher than in bulk metal; the exact value depends on the deposition process and the resulting film stoichiometry, crystallinity, and purity. For larger displays, some of the key parameters are the busline resistances, which need to be low enough to minimize signal propagation delays and resulting distortions of address pulses. The resistance R of a busline of metal film is

$$R = \rho \frac{l}{wd},\qquad(3.1)$$

where l is the length of the busline, w its width, and d the thickness of the film.

A useful parameter is the sheet resistance, defined as

$$R_{sheet} = \frac{\rho}{d}.\qquad(3.2)$$

Table 3.3: Resistivities of thin film materials used in AMLCDs

Material	Function	Resistivity ($\mu\Omega$cm)
Ta	Gate electrode, select busline	20–200
Cr	Gate electrode, select busline	20
Mo	Source/drain electrode, data busline	13
Al	Both electrode, both buslines	4
Cu	Gate electrode, select busline	2.5
Al alloys	Gate electrode, select busline	4–10
ITO	Pixel electrode	100–300

It can be viewed as the resistance of a square section of the film (i.e., $w = l$) and its unit of measure is the ohm. Since sheet resistance applies to a square section of film, it is also often expressed as ohms per square ($\Omega/\square$).

The total resistance of a busline can therefore be easily calculated from Eqs. 3.1 and 3.2 as

$$R = R_{sheet}\frac{l}{w}. \qquad (3.3)$$

For large displays (>15 in.) it is necessary to use low-resistivity metals such as Al or Cu.

For dielectrics, used as the gate insulator and in the storage capacitor, an important parameter is the capacitance per unit area C_{unit}:

$$C_{unit} = \frac{\varepsilon_0\varepsilon}{d}, \qquad (3.4)$$

where ε_0 is the permittivity of vacuum (8.85E-14 F/cm), ε is the dielectric constant, and d is the thickness of the film. Table 3.4 lists the dielectric constants and refractive indices of some materials used in AMLCDs.

The leakage current of the insulators should be small enough to not affect the display operation at refresh rates around 60 Hz. The important parameter is the RC relaxation time of the capacitor. It determines the decay of the voltage over time according to the formula $V = V_0 exp(-t/RC)$. Since, for a capacitor with area A and thickness d, $R = \rho d/A$ and $C = \varepsilon\varepsilon_0 A/d$, it follows that

$$RC = \rho\varepsilon\varepsilon_0, \qquad (3.5)$$

Table 3.4: Dielectric constants and refractive indices of materials used in AMLCDs

Material	Function	Dielectric constant	Refractive index
SiO$_2$	Gate insulator, passivation	3.9	1.5
Si$_3$N4	Gate insulator, passivation	6–8	1.8
LC fluid	Light modulation	3–14	1.5
Color filters	Color generation	4.5	1.5
Polymer overcoats	Passivation, planarization	3.5–5	1.5
a-Si	Semiconductor	12	3.4
ITO	Pixel electrode		1.8–2.0
Polarizers	Light modulation		1.5
Glass	Substrate		1.5

where ρ is the resistivity of the insulator (>1E12 Ωcm). In other words, the RC relaxation time is a material parameter and is independent of the area A of the capacitor and the thickness d of the capacitor dielectric. Equation 3.5 also applies to the LC capacitance. As discussed in Chapter 2, Sec. 2.2, the voltage holding ratio depends on the leakage through the LC. The voltage decay on the LC capacitance during one frame time is given by

$$V_{LC}(t) = V_{LC}(0)\, exp\left(\frac{-t}{RC}\right). \tag{3.6}$$

Equation 3.5 implies that the voltage holding ratio is a material parameter, independent of the LC cell gap and pixel area.

The semiconductor film's function is to modulate the current in the TFT, depending on the voltage applied to the gate. The field effect mobility, threshold voltage, and OFF current of the TFT depend to a large degree on the quality of the semiconductor film.

Important parameters other than electrical properties are the mechanical and optical properties of the films.

Mechanical properties of interest include film stress and morphology. Stress needs to be minimized to avoid film adhesion loss, cracking, and buckling of the films and bending of thin substrates. Morphology includes the crystallinity and other growth-related features, such as columnar structure. Columnar growth of some metals can increase the probability of stress fractures in the film at steps, leading to open buslines. For example, molybdenum (Mo) can be used as the data busline metal, but its columnar growth can cause the busline to crack open at the step of the data busline over the gate line.

Aluminum (Al) is a widely used busline material for gate lines and source lines. It is much more ductile than the refractory metals and therefore is less susceptible to line opens. However, a serious mechanical issue with aluminum is its tendency to grow hillocks during high-temperature processing steps. They are the result of film buckling at high temperatures related to the differences in thermal expansion between glass and aluminum. Hillocks cause serious yield problems by shorting the Al layer to subsequently processed conductors. This issue is addressed by alloying aluminum with rare earth metals, such as Nd. A thin cap layer of Ti or other metals on top of the aluminum layer can also effectively suppress hillock formation.

Optical film properties are obviously very important for the transparent conductor layers and for the color filters and black matrix layer. The absorption coefficient and refractive index are the key parameters. Chromaticity for the color filters will be discussed in Chapter 5. The transmittance T of a film with thickness d and absorption coefficient α is given by

$$T = (1 - R)\exp(-\alpha d),\qquad(3.7)$$

where R is the film reflectance. T, R, and α are, in general, all wavelength-dependent. For very thin films, Eq. 3.7 is just an approximation because interference effects in the film can cause peaks and valleys in the transmittance and reflectance spectra.

Display brightness is proportional to the transmittance of the transparent conductors. The ITO films used for the pixel electrodes on the active matrix array and the color filter plate typically have a transmittance of more than 90% for the entire visible spectrum from 400–700 nm.

The reflectance R at any interface for normally incident light between two low-absorbing materials in the LCD is given by

$$R = \frac{n_1 - n_2}{n_1 + n_2},\qquad(3.8)$$

where n_1 and n_2 are the refractive indices of the two materials. Since there are many different layers in the optical path through the LCD (polarizers, glass, LC fluid, ITO electrodes, color filters, etc.), the concept of index matching is important to minimize losses from reflectance at the various interfaces. Table 3.4 lists the refractive indices of some materials used in AMLCDs. Most are close to 1.5, so that reflectance at their interfaces is small.

The black matrix material used to block light in inter-pixel areas needs to be opaque. The opaqueness of a thin film can be expressed in its optical density OD:

$$OD = {}^{10}\log\left(\frac{I_0 - RI_0}{I}\right),\qquad(3.9)$$

where I_0 is the incident light intensity on the film and I is the transmitted light intensity. It follows from Eq. 3.7 that the optical density is proportional to the film's thickness d and its absorption coefficient α. For metal black matrix films it is easy to achieve a high OD because of the high α of most metals. For dark polymer resin, black matrix layers α is smaller and a film thickness exceeding 1 μm may be needed to obtain an $OD > 2$ or 3 needed to get high-contrast-ratio displays.

All parameters described in this section are considered in the design optimization of the TFT array and color filter array for the LCD. In addition, these parameters need to be uniform across the entire display area.

3.4   Amorphous Silicon TFT Array Processes

The displays in notebook computers, flat panel monitors, and LCD televisions are almost exclusively amorphous silicon (a-Si) TFT LCDs. A close look at their manufacturing process is therefore warranted.

In Figs. 3.14 and 3.15 the process steps for a typical a-Si TFT pixel array are outlined step-by-step. The pixel layout, as it evolves during the process, is shown in the figures along with the cross section along the line CC′.

The starting substrate is 0.7-mm Corning Eagle®, Corning 1737®, or equivalent glass. They are fused borosilicate glass substrates with low sodium content and have been specially developed for display applications. There has been a migration from glass with thickness of 1.1 mm during most of the 1990s to the current preference for 0.7 mm to reduce panel weight. Some TFT LCD factories have transitioned to 0.5-mm glass for notebook and smaller panels.

First, a 200–300 nm layer of gate metal is deposited by sputtering and is subsequently patterned. This first metal layer also functions as the select buslines in the display. Aluminum or an Al alloy is the most common metal used. To improve adhesion of the Al film to the glass, it can be preceded by a thin refractory metal layer such as Ti. To prevent buckling of the Al film during subsequent higher-temperature processing (the hillock formation discussed in the previous section), the Al is often capped by another thin refractory metal layer such as Mo or Ti. After defining the photoresist with the first mask, the layer is patterned by wet etching or dry etching.

Secondly, the gate dielectric of silicon-nitride (SiN), the intrinsic a-Si layer, and the n^+ a-Si layer are sequentially deposited by PECVD without breaking the vacuum. Their thicknesses are typically 300–400 nm, 150 nm, and 20 nm, respectively. These three layers used to be deposited in separate chambers of the PECVD system. To increase throughput they are now commonly grown in the same chamber, interrupted between layers by brief flushing procedures to change the gas mixture. The second mask step is used to define the Si island. The selective dry etching step to achieve this etches only the n^+ a-Si and intrinsic a-Si layers and leaves the gate SiN intact as a blanket film covering the entire substrate.

Next is the deposition of the source-drain metal by sputtering. This can be the same metal (for example, 200–300 nm of Ti-Al-Ti or Mo-Al) as used for the gate metal to minimize the number of metallization recipes in the production line. The third mask delineates the source and drain of the TFT along with the data busline. The Ti in

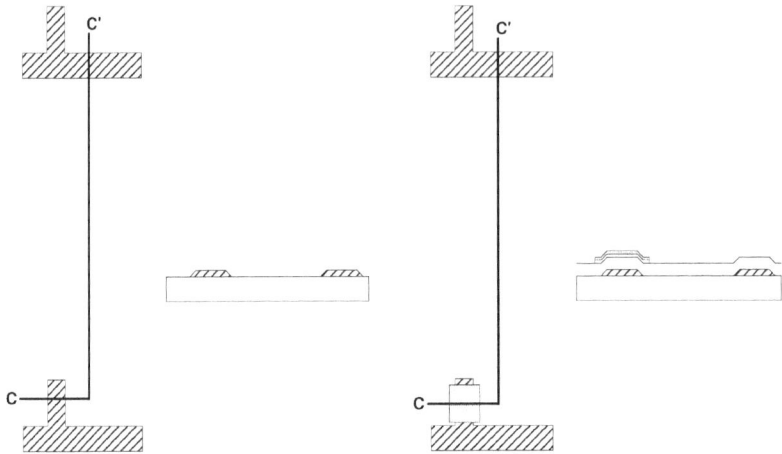

1. Gate metal deposition and patterning

2. SiN/a-Si/n+ deposition, a-Si patterning

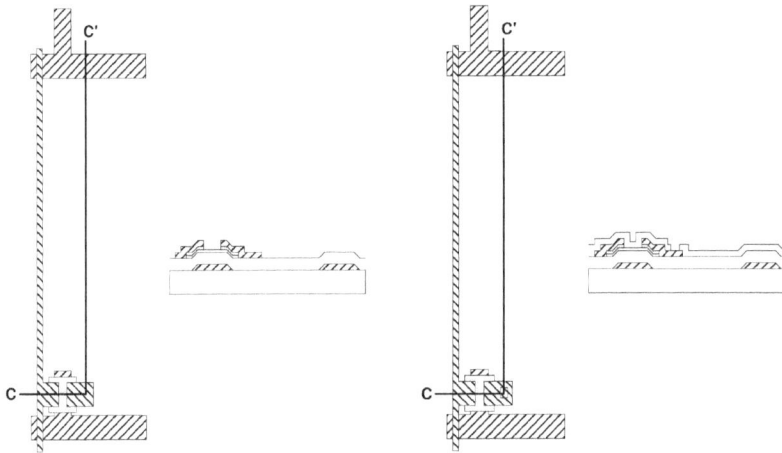

3. Source/drain metal deposition and patterning

4. Passivation deposition and patterning

Figure 3.14: Layout and cross section CC′ of a-Si TFT subpixel, as it evolves during the first four mask steps.

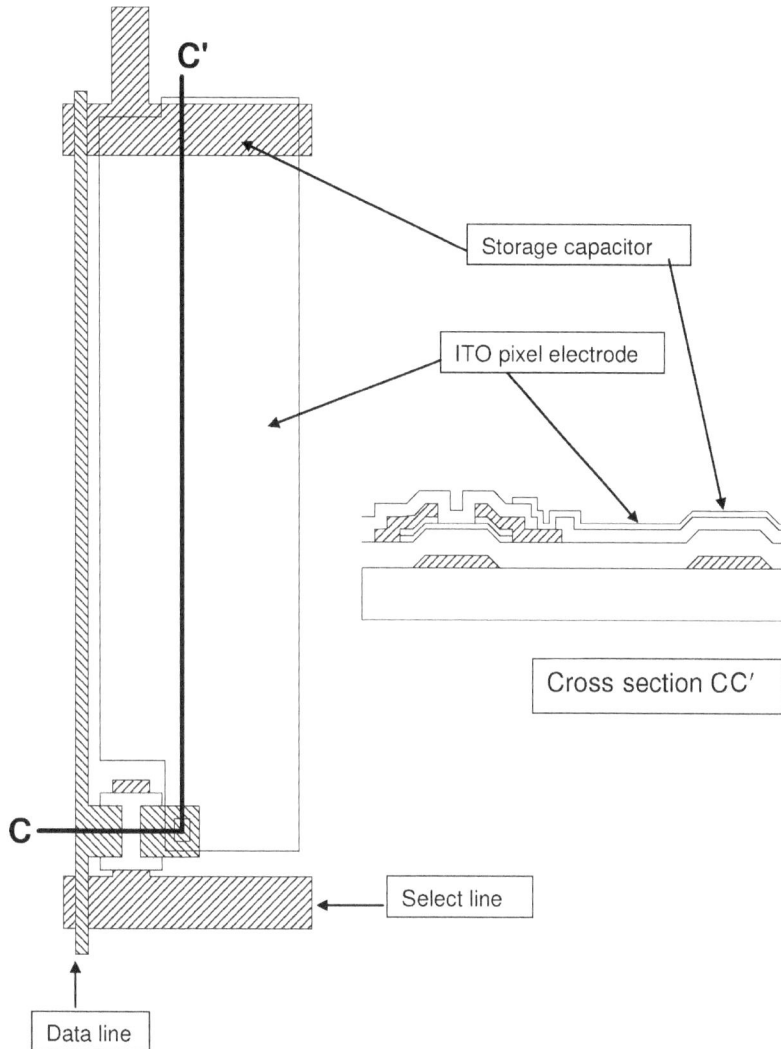

Figure 3.15: Pixel layout and cross section after the fifth and final patterning step of the ITO pixel electrode.

direct contact with the n$^+$ layer provides a good, low-resistance ohmic contact for the source and drain.

After the source and drain are patterned, the relatively conductive a-Si n$^+$ layer is still present in the channel, causing a high OFF current in the TFT. The n$^+$ layer is removed from the TFT channel by a dry etch step called the back-channel etch. This dry-etch step does not selectively etch the n$^+$ and intrinsic a-Si layer (i.e., both have about the same etch rate). The back-etch, therefore, needs to be sufficiently uniform to remove the n$^+$ layer completely across the entire substrate without removing too much of the intrinsic a-Si film, so that enough silicon is left in the channel to obtain a well-functioning TFT with sufficient ON current.

Next is the deposition of an approximately 200-nm passivation silicon-nitride layer to protect the TFT and buslines from the LC fluid. The passivation layer is patterned with the fourth mask to make openings (vias or contact holes) at the TFT drain locations and at the periphery of the display to make interconnections.

Finally, an ITO layer with a thickness of 300–700 nm is deposited and patterned into pixel electrodes with the fifth mask. The ITO pixel electrode makes contact to the drain of the TFT through the contact hole in the passivation SiN layer.

Figure 3.15 shows the final pixel layout and cross section.

The sequence of the process steps can vary from factory to factory. In some manufacturing lines the ITO layer is deposited and patterned prior to the source and drain metal. In many factories the ITO transparent pixel electrode is now deposited and patterned as the last step after passivation of the array to minimize the possibility of shorts between the data buslines and ITO pixel electrodes. Some plants use a trilayer TFT rather than a BCE TFT (see Sec. 2.5 in Chapter 2) at the expense of one more PECVD and one more photomask step.

The initial notebook panel products on the market in the 1990s typically used six to eight mask processes for the TFT array, adding, for example, an extra gate metal patterned separately, or adding a separate step for etching the gate nitride, or using a trilayer TFT. Each photolithography step adds significantly to the overall process cost. Many efforts have therefore been undertaken to reduce the total mask count, with the five-mask process described above as the most common result.

A four-mask process has been implemented in some fabs by combining the patterning of the a-Si and the source-drain metal in a single photolithography step. This is done by using a special mask that allows for partial diffractive exposure of the photoresist in the channel region of the a-Si TFT (Fig. 3.16). When the photoresist is subsequently etched back, it can be removed from the channel area while still covering the source/drain contacts and data buslines. The source/drain metal and n$^+$ layer can then be etched away from the channel area. Although this process eliminates one mask and photolithography step, it needs an additional dry-etch step to etch back the photoresist, a duplicate etching step for the source-drain metal, and a unique photomask that is difficult to manufacture. Such simplifications can lead to cost reductions, but sometimes also require tighter process control to maintain high yield.

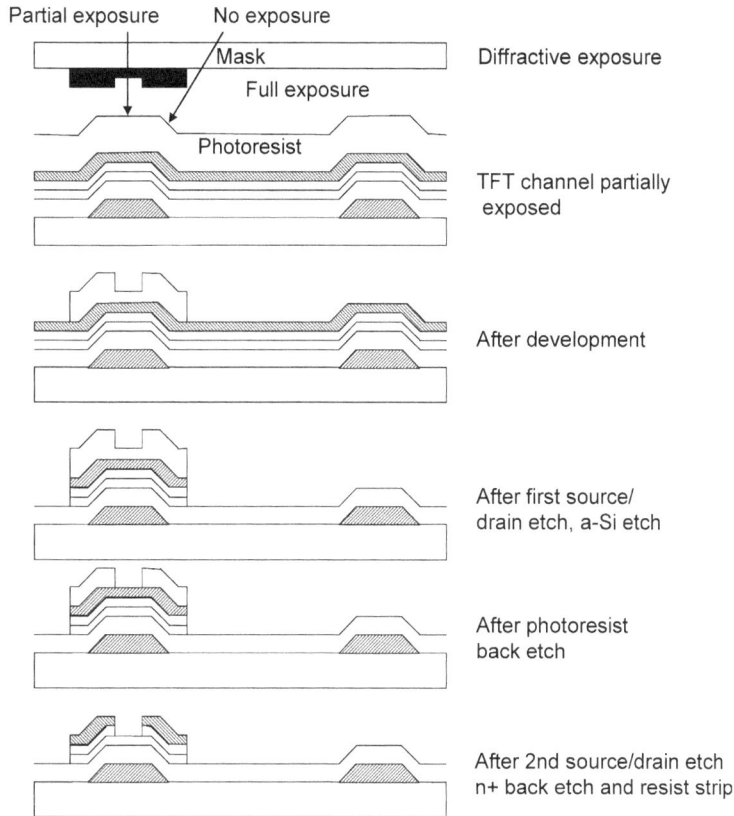

Figure 3.16: Combination of Si island and source-drain patterning in one step by diffractive exposure.

3.5   Poly-Si TFT Array Processes

Poly-Si-based LCDs need more process steps than a-Si TFT LCDs. They are mostly used in smaller displays where the integration of row and column drivers on the glass cancels out the higher processing cost.

High-temperature poly-Si LCDs are applied in projection displays and their TFT arrays are manufactured on quartz substrates with processing mostly borrowed from the semiconductor industry. The Si film is crystallized by solid phase crystallization (a high-temperature annealing step) after deposition at 900–1100°C. Following steps include thermal oxidation to obtain a gate SiO_2 film and ion implantation for doping the n^+ and p^+ contacts. More advanced design rules than are common in direct-view displays are used to obtain the small features for projection displays with integrated peripheral electronics.

Low-temperature poly-Si (LTPS) TFTs are fabricated on glass substrates with all steps below 600°C, and include custom processing steps for crystallization, gate insulator deposition, and doping. Solid phase crystallization must be replaced by a lower-temperature crystallization procedure.

In Fig. 3.17 an example of a basic process sequence for producing LTPS TFT arrays is outlined. Key steps are the laser crystallization of a-Si into poly-Si, the gate insulator process, and the ion doping for the source/drain contacts. The starting substrate is glass such as Corning Eagle® or Corning 1737®. Although this type of glass has a low sodium content, it usually needs to be coated first by SiN and SiO_2 blocking layers to prevent any remaining sodium to diffuse into the TFT structures at 600°C and affect the threshold voltage and stability of the LTPS TFT. The a-Si film is deposited first by PECVD and then crystallized.

The poly-Si is subsequently patterned into islands and a gate insulator film of SiO_2 is deposited. The gate electrode deposition and patterning follow. Then, two large area ion doping steps are used to obtain p^+ and n^+ contacts for the source and drain. To achieve a low transistor OFF current, an additional doping step may be added prior to p^+ and n^+ doping to produce a lightly doped drain (LDD) structure. The dopants are activated by annealing steps. The doping process is followed by field oxide deposition and patterning of vias (contact holes). Then the source/drain metal is deposited, which contacts the p^+

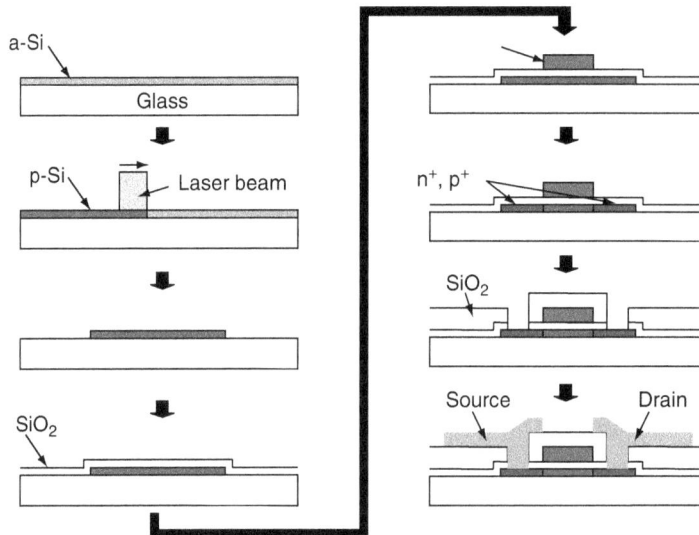

Figure 3.17: Basic sequence of a poly-Si TFT process.

and n$^+$ areas through the vias and also functions as the low-resistance buslines. In the viewing area of the display, two additional steps (not shown in Fig. 3.17) form the passivation layer and the ITO pixel electrodes.

The entire process typically requires about eight photolithography steps, significantly more than a-Si TFT arrays. Some manufacturers use only p-type poly-Si TFTs to simplify the process and to obtain a lower TFT OFF current. This requires the implementation of peripheral electronics with PMOS rather than CMOS circuits.

Much effort has gone into optimizing the crystallization process, which determines, to a great extent, the grain size of the poly-Si and the TFT field effect mobility. A number of companies use the excimer laser annealing (ELA) technique shown in Fig. 3.18.

Ultraviolet pulses with a wavelength of 308 nm from the 200-W XeCl laser are strongly absorbed by the thin Si film. The Si layer melts within 5 ns at about 1400°C without raising the glass temperature beyond its softening point of 700°C. Upon cooling down, the Si layer is recrystallized as poly-Si with a grain size depending on the original Si film thickness and morphology, the laser energy, the speed of scanning, and other parameters.

Other crystallization procedures include the continuous grain silicon (CGS) method developed by Semiconductor Energy Labs and Sharp Corporation. It uses a seed layer to enhance the crystallization and increase grain size and film quality. Lateral growth methods in which crystallization starts at a localized site and progresses laterally along the film direction have also been developed.

Doping steps are used for the source-drain contacts, for the lightly doped drain to reduce OFF current, and often also in the channel to reduce the threshold voltage for

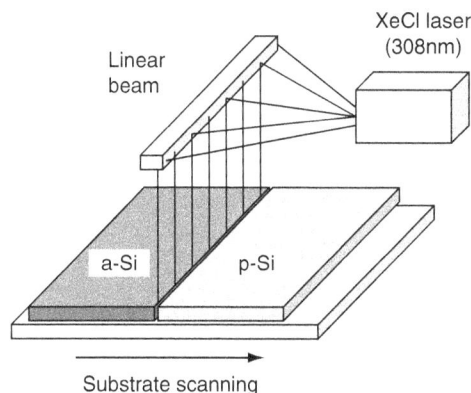

Figure 3.18: Crystallization by excimer laser annealing (ELA).

both P-channel and N-channel TFTs. The lowering of the TFT threshold voltage is needed to ensure low operating voltages for CMOS circuitry integrated at the periphery of the display.

The ion-doping process allows relatively large areas to be doped at much higher speed than is possible with conventional ion implantation. The equipment consists of a plasma generating source with a feed gas of B_2H_6 (for p-type doping with boron) or PH_3 (for n-type doping with phosphorus). The resulting species in the plasma are accelerated by high-voltage electrodes without mass-separation and are implanted into the silicon film. The dopants are subsequently activated by one of three possible annealing steps: furnace annealing, another ELA step, or rapid thermal annealing with a high-intensity light source.

The SiO_2 gate insulator for LTPS TFTs cannot be obtained by thermal oxidation of the Si film because of the temperature processing limit of 600°C. Several deposition techniques can be used instead to approach the quality of thermal oxides. The interface between the Si and SiO_2 needs be ultra-clean to avoid interface states. Plasma CVD is successfully being used commercially for gate oxide deposition. An example of a process that gives stable TFTs is PECVD in a mixture of O_2 and $Si(OC_2H_5)$, also called TEOS for tetraethooxysilane, deposited at 400°C.

The LTPS process has many variables, which explains the widely varying results obtained by different organizations and companies. This is evident from Table 3.5, showing which process steps have an influence on the mobility, threshold voltage, uniformity, and stability of the LTPS TFT. In comparison, the a-Si TFT process has fewer variables, and different groups generally report similar results for mobility and threshold voltage.

Table 3.5: Process steps affecting LTPS TFT parameters

Step	Mobility	Vth	Uniformity	Stability
Protective blocking layer		yes		yes
Si deposition	yes			yes
Si crystallization	yes	yes	yes	
Doping of channel		yes	yes	
Gate insulator deposition		yes		yes
LDD, source/drain doping			yes	yes

3.6 Color Filter Array Process

Color filter plates are often manufactured by different companies than the TFT arrays and are then shipped to the LCD maker for panel assembly. Some AMLCD manufacturers have developed in-house capability to manufacture color filter plates to eliminate shipping and reduce their dependence on outside vendors.

The color filter array process (Fig. 3.19) starts with the deposition and patterning of a black matrix layer. Its purpose is to block light in the inter-pixel areas and shield the TFT channel from ambient light. The black matrix layer can be a metal such as chromium, often with a chromium oxide layer added between the chromium and the glass to reduce display reflectance. The Cr/CrO_2 combination is often referred to as black chrome and helps with maintaining contrast ratio at high ambient lighting levels. Because of environmental concerns with the use of Cr, it has in some cases been replaced by a black polymer resin. The black polymer can be photo-imageable and is therefore easier to process than the metal black matrix. Another feature of the black polymer resin is a

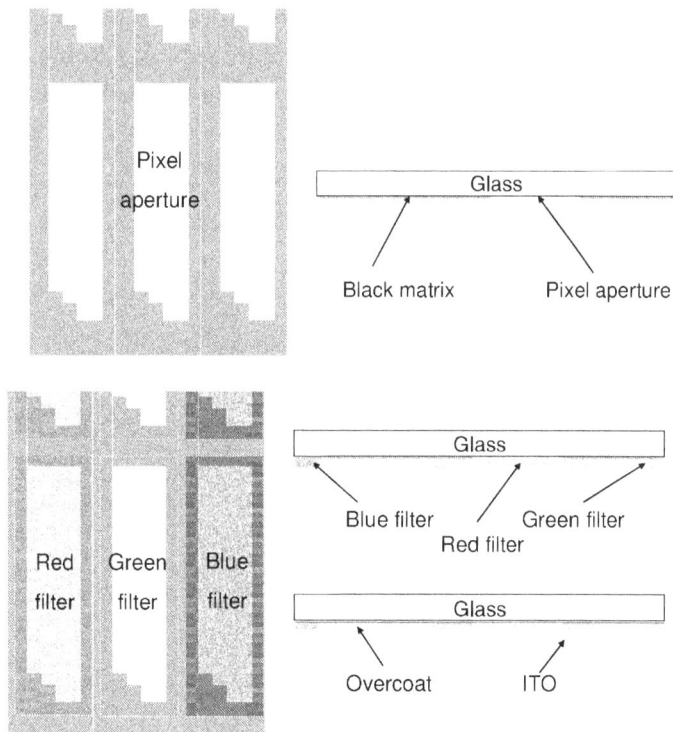

Figure 3.19: Color filter array process: pixel layout and cross section.

73

refractive index close to that of glass. This minimizes reflectance of ambient light by the black matrix.

After the black matrix process, red, green, and blue color filters are deposited and patterned. Different types of color filters can be used, either based on color dyes or on color pigments. In the simplest process (now commonly used), the color filter material has dispersed pigments and has added photo-initiators. It is therefore photo-imageable by itself under UV exposure, so that photoresist coating and stripping steps are eliminated. Optionally, a transparent overcoat layer may be deposited to planarize the surface on top of the color filters. As the last step, the common transparent ITO electrode is deposited. This is, in most cases, a blanket layer covering the entire viewing area. It may be kept off the peripheral area by the use of a shadow mask during sputter deposition.

3.7   LC Cell Assembly

After the TFT array and color filter plates have been processed, the LC assembly process joining the two substrates starts (Fig. 3.20).

Each plate is coated with a polyimide layer and rubbed to create a preferred alignment direction of the LC molecules at the two surfaces. The rubbing or buffing is done by a rotating drum covered with fine-haired cloth contacting the substrates (Fig. 3.21). This

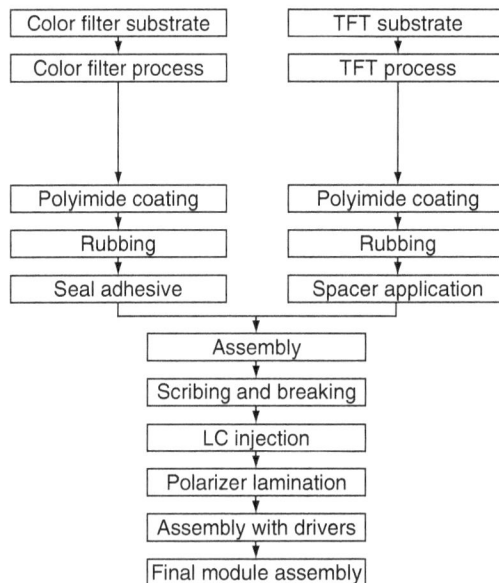

Figure 3.20: Process sequence in LC and module assembly.

74

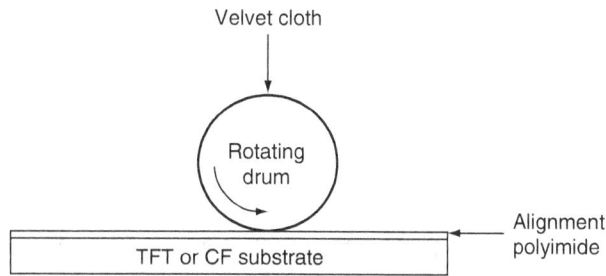

Figure 3.21: Rubbing process to create preferential alignment of the LC molecules at the substrate surfaces.

critical process step determines the alignment direction of the LC molecules at the surface, and their anchoring strength. Parameters such pressure, rotating speed, and quality of the rubbing cloth play an important role and also affect the pre-tilt angle of the LC molecules at the surface. The rubbing process step has often been optimized empirically with extensive experimentation. It remains a source of yield loss, causing non-uniformities in the dark state of the LCD. Some LC modes, such as the MVA and PVA modes to be discussed in Chapter 6, do not require a rubbing process, which is considered a yield advantage.

Subsequently, spacers (plastic spheres with a diameter of 4–5 μm) are sprayed on the TFT plate. Their function is to accurately control the LC cell gap between the two plates. The peripheral glue seal is dispensed on the color plate, along with conductive epoxy dots in the corners of the color plate. The conductive dots will serve to short the ITO common electrode on the color plate to metal pads on the TFT array plate, so that the V_{com} voltage can be applied to the color plate through pads on the TFT array plate.

This is followed by the assembly of the two plates with a plate-to-plate alignment accuracy of better than about 3 μm. An injection hole or fill port is left open in the glue seal to allow filling of the assembly with LC fluid. The filling process takes place in a vacuum chamber. A stack of LC assemblies is evacuated in a vacuum chamber to a pressure below 1 Torr (Fig. 3.22). The glue seal openings are then dipped into LC fluid and the chamber is backfilled with N_2 gas. This forces the LC fluid to enter the narrow cell in a process that can take, for large displays, many hours to complete.

In Fig. 3.23 the combined layout and cross section of color filter pixels and TFT array pixels after assembly of the two substrates are shown. The black matrix opening is, in conventional designs, about 4–8 μm smaller on each side than the ITO pixel electrode to allow for plate-to-plate misalignment and to prevent light leakage at pixel edges.

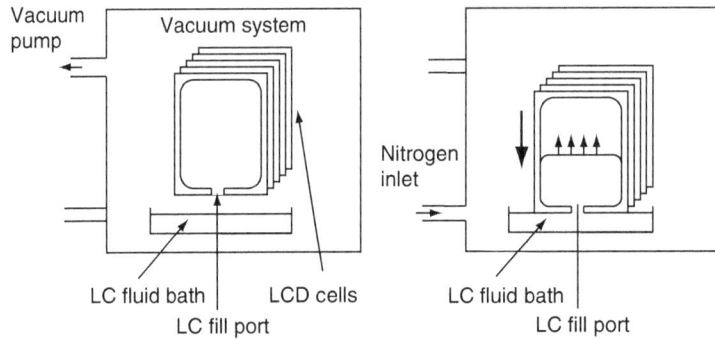

Figure 3.22: Filling of LC cells through fill ports by nitrogen backfill in a vacuum system.

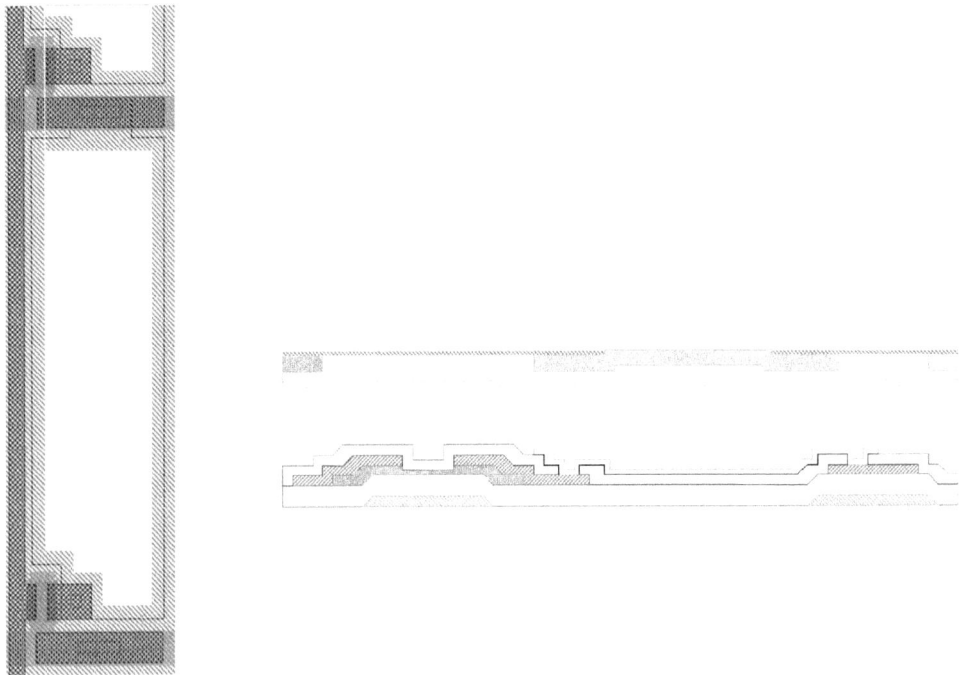

Figure 3.23: Combination of active array and color plate.

Recently, several new assembly techniques have been introduced in manufacturing (Fig. 3.24). One is the use of photolithographically patterned spacers to better control the cell gap and the location of the spacers. These so-called column spacers are patterned outside the pixel aperture area. Therefore, they have no impact on contrast ratio. On the contrary, randomly sprayed plastic spacers end up in the pixel aperture area, generating

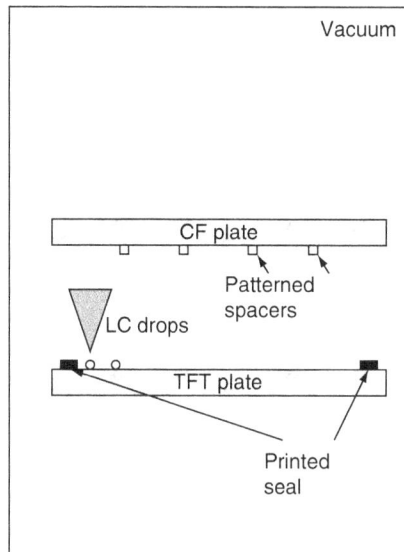

**Figure 3.24: Newer LC assembly
techniques using patterned spacers
and LC drop filling.**

light leakage around them. Contrast ratios exceeding 1000:1 have been reported with patterned spacers. Another process improvement is plate-to-plate assembly in vacuum after drops of LC are dispersed on one of the substrates. The drop-filling process provides a dramatic reduction in the filling time from more than 24 hours to a few minutes for large displays used in LCD televisions and monitors.

As mentioned before, the rubbing process to create the orientation layer is sometimes a yield detractor. One company has replaced the polyimide alignment layer with a diamond-like carbon (DLC) layer, which is treated with an ion beam process to obtain the preferential surface alignment for the LC molecules. By eliminating the rubbing process, a better manufacturing yield is possible with the DLC process. Other orientation methods based on photo-alignment (exposure to UV light with a preferred polarization or direction) have been developed as well.

3.8 Module Assembly

The next step is the module assembly, which starts with the polarizer lamination. Most LCDs require a polarizer to be attached on each side of the glass assembly. This is followed by the driver attachment by either tape automated bonding (TAB) or by a chip-on-glass (COG) process.

In the TAB process, the tape carrier packages (TCPs) containing the row and column drivers are bonded to the edge of the glass (Fig. 3.25). The TCP has inner lead bonds and outer lead bonds. The inner lead bonds connect the chip inputs and outputs to the flex. The outer lead bonds connect the outputs on the flex to the display glass and the inputs on the flex to the row board and to the column board, which contains the video interface, timing controller, and other circuitry (Fig. 3.25). The TAB bonding process is done by pressing a heated bar on the sandwich of the tape, adhesive, and glass.

An anisotropic conductive adhesive is used, which consists of small plastic spheres with a conductive coating suspended in a glue material. A heat seal is formed between the glass and the TCP with the conductive particles in the glue contacting the conductors on the glass with those on the TCP. Since the conductive spheres are small, they conduct only in the vertical direction without causing shorts between adjacent leads on the flex or the glass, hence the name anisotropic conductive adhesive.

In the COG process (Fig. 3.26), the flex is eliminated by bonding the driver chip directly upside-down on the glass, also with an anisotropic conductive adhesive. The COG process requires more routing of the signals, including chip input signals, on the edge of the display glass. COG was originally mostly used on smaller displays and can reduce cost by eliminating the tape. Some manufacturers now use it even on 17-in. panels.

Finally, the module assembly is completed by adding backlight, diffuser, and light guide (optional) in a housing (Fig. 3.27). The module assembly process is quite labor-intensive and has, in many cases, been moved to lower-labor-cost countries such as China.

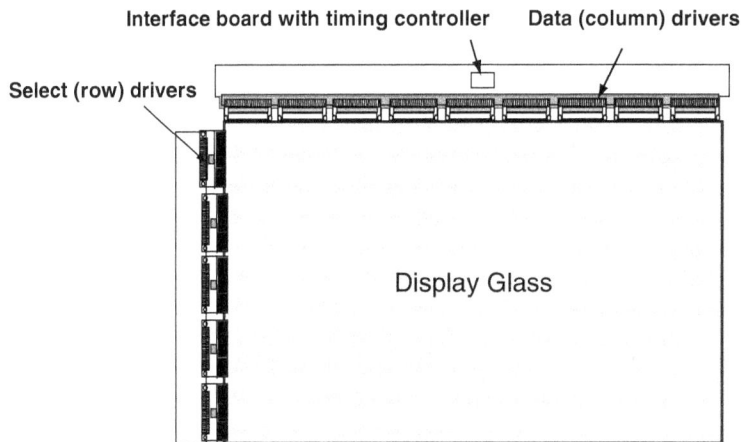

Figure 3.25: Assembly with drivers.

Anisotropic conductive adhesive

Driver chip

Color plate

TFTplate

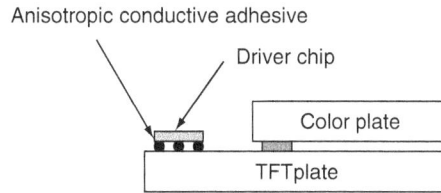

Figure 3.26: Driver connection to the glass with chip-on-glass (COG).

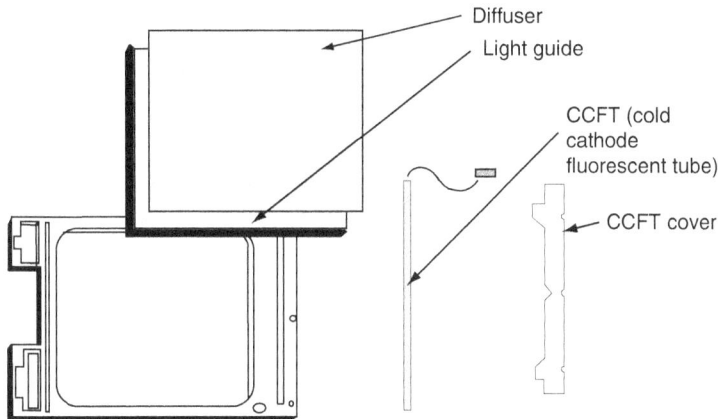

Diffuser

Light guide

CCFT (cold cathode fluorescent tube)

CCFT cover

Figure 3.27: Final module assembly.

3.9   Yield Improvements and Considerations

A closely guarded secret of most LCD panel manufacturers is their manufacturing yield, for obvious competitive reasons. Yield improvements have been dramatic due to the introduction of automation and by improvement of processes, materials, and component quality. Yield depends on the specifications. In the past, a number of pixel defects were allowed in notebook displays and other products. The acceptable defects are clearly defined in documents to minimize the potential for conflicts between panel suppliers and customers. Over the years, with the improvements in manufacturing, the number of acceptable pixel defects has been reduced, in many cases to zero. Non-uniformities and variations in contrast, brightness, viewing angles, and color performance in the final product have also been reduced.

A high yield in the front-end process starts with state-of-the-art facilities with Class 10 clean rooms (less than 10 particles per cubic foot), proven manufacturing equipment and processes, and full automation without manual handling of the substrates at any time. The large manufacturing substrates (around 3–4 m^2 for Generation 6 and 7 factories) are transported in large cassettes between process steps. When filled with 10 to 20 substrates, the cassette can weigh more than 500 kg, requiring large, powerful robots.

Reduction of particles in the manufacturing environment is one of the keys to high yield. As in semiconductor manufacturing, the yield Y depends on defect density according to the following generic equation:

$$Y = \exp\{-A \sum_{n=1}^{N} D_n\},$$ (3.10)

where A is the area of the display, D_n is the density of killer defects introduced at step n, which cannot be repaired, and N is the number of process steps. The yield decreases monotonically as the display area increases.

Yield is also optimized by design for manufacturing: the tFT array design rules are set up with an eye on maximum possible process latitude. The yield is constantly monitored and yield losses are categorized according to their highest frequency occurrence in Pareto charts. Key process parameters in all process steps are monitored and acted upon continuously using statistical process control (SPC) methods. They include the so-called critical dimensions for the patterning (layer-to-layer overlay and line width and spacing), thicknesses, and uniformity of all layers, including photoresist, process temperatures, and many of the equipment variables.

Any jumps in yield are correlated, if possible, to a change in one of the key parameters. If a process is out of statistical control, corrective action needs to be taken to bring it back into the fold. Regularly scheduled preventive maintenance on the equipment helps keep processes within their acceptable window.

Often, special circuitry is added in the array design to protect the TFT arrays in the viewing area against electrostatic damage (ESD). ESD can be an issue at different steps in the manufacturing process, such as during the transfer of substrates between equipment and transportation cassettes.

An example of ESD protection circuitry added at the periphery of the display is shown in Fig. 3.28. It consists of several TFTs with their gates shorted to the source and starts conducting current when excess voltages of positive or negative polarity appear between the data lines and select lines. In Fig. 3.28 each protection circuit consists of four TFTs, two in series in each branch to prevent high transient voltages of either polarity from reaching the viewing area. More than one TFT in each branch is preferred to increase the threshold for conduction, so that no excess leakage occurs between the select and data lines during normal operation of the display. This kind of circuit helps prevent undesirable TFT threshold voltage shifts in the viewing area by absorbing the effect of high transient voltages that can occur during the manufacturing process.

At several steps during the TFT array manufacturing process, the arrays are inspected and tested. For example, the photolithographic patterning can be checked with automated optical inspection (AOI) equipment, which scans the substrates at high speed for patterning defects using high-resolution CCD cameras and pattern-recognition software. AOI can detect defects in transparent conductor (ITO) patterns as well as in a-Si and metal patterns. The patterning defects can be reviewed, classified, and acted upon if necessary. When potential killer defects are detected after photoresist development but prior to etching, the substrates can be reworked by stripping the resist and repeating the coat and exposure steps. When the array process is finished, a functional test is performed that gives the number and location of pixel defects and marginally performing pixels, if any. One type of equipment does this by connecting probes to all leads and charging up the storage capacitor at each pixel by scanning all select lines and applying uniform data to the source lines, in a fashion similar to actual display operation. In defective and weak pixels, the charge on the pixel capacitor will not be sustained. Subsequently, the array is scanned again and the charge on each pixel is read out by a detection circuit. This method indicates which pixels are defective or have inaccurate gray levels on the TFT array. The functional test of the array prior to assembly with the color plate prevents the

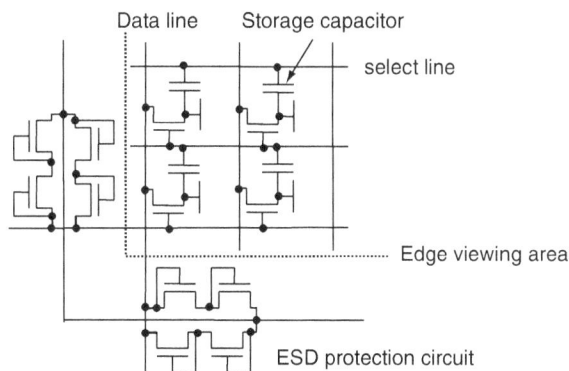

Figure 3.28: Example of ESD protection circuitry added at the edges of the viewing area in TFT LCDs to prevent ESD damage.

combination of non-yielding active plates with costly color filter plates. As inspection and testing by themselves do not improve yield, their purpose is also to find out where yield losses occur and then implement corrective action or a repair procedure.

Display specifications allow fewer and fewer, if any, pixel defects. They can be categorized as permanently bright or permanently dark pixels. Bright pixel defects are even less acceptable, especially in large displays with large pixels, such as LCD televisions, because they stand out and are very visible. In some cases, bright subpixel defects can be converted into less visible, permanently dark subpixels after finding their location in the functional test of the TFT array. For example, a shorted TFT can be disconnected from the subpixel ITO electrode by laser zapping. This causes the subpixel electrode to float at all times, so that no voltage can be applied across the pixel. For a normally black LCD, this would make the subpixel permanently dark.

Repair of short or open buslines by laser methods is also common. For example, a finished large TFT array with an open source line or open gate line already has a large added value after going through the entire process. To reduce scrap and save costs, it is attractive to perform a repair operation, as shown in Fig. 3.29. The open line, which would result in a visible, partially open line defect in the display, is connected to added repair lines at the periphery of the display by laser zapping. The repair lines are routed around the display on the glass edge and/or the printed circuit boards, to be connected to the original signal line or to extra outputs on the driver chip specifically added for this purpose.

The farther along the AMLCD is in the production process, the larger its value. In back-end processing, any yield loss is therefore considered unacceptable.

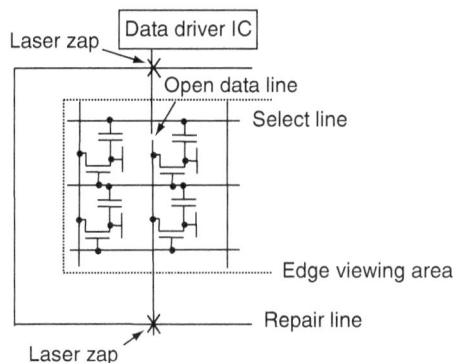

Figure 3.29: Example of repair of open buslines in a TFT LCD.

3.10   Trends in Manufacturing

The manufacturing of a-Si TFT arrays progressed through the 1990s from mother glass sizes of 350×400 mm (Generation 1) to 730×920 mm (Generation 4). Larger mother glass substrates increase productivity, since more unit displays can be fitted on a substrate. This trend continues during this decade (Fig. 3.30). Figure 3.31 shows how different-sized LCD televisions can be fitted on Generation 5, 6, and 7 substrates.

An increasing number of Generation 5 factories (1000×1200 to 1100×1300 mm) and Generation 6 factories (1500×1800 mm) are currently in operation. Figure 3.32 gives an impression of the size of the substrate at the first Generation 5 fab at LG Philips LCD in Korea. The substrate can fit nine 18.1-in. displays or fifteen 15-in. displays. In the

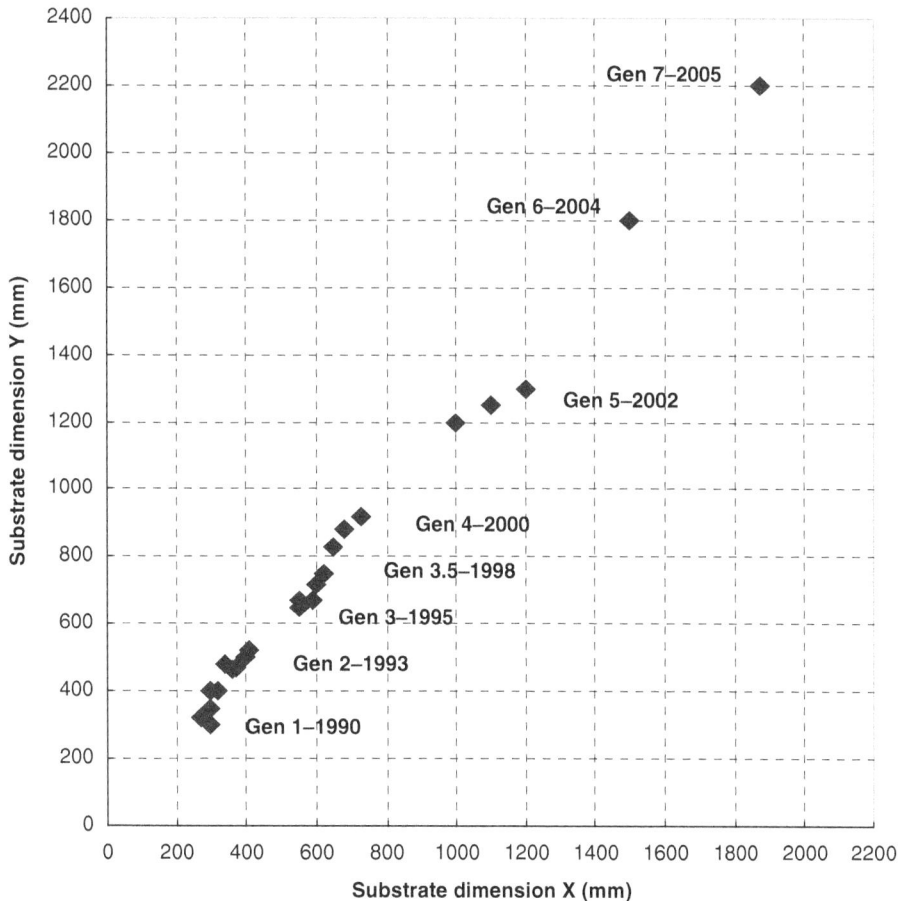

Figure 3.30: Evolution of substrate sizes in different Generation factories.

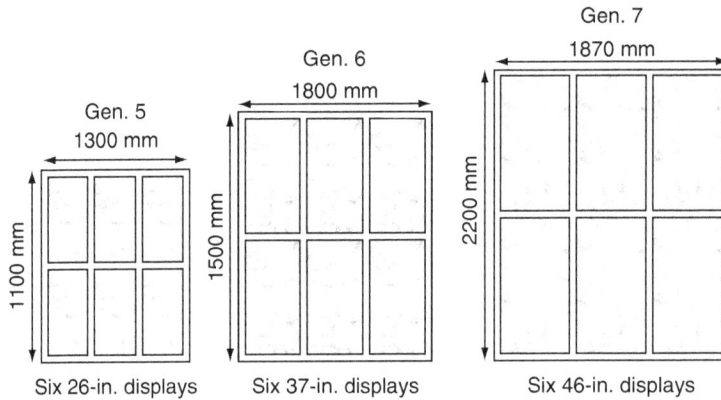

Figure 3.31: Optimized TV display sizes on different-sized substrates.

photograph, two operators are holding the substrate for demonstration purposes only. The factories are fully automated and substrates are not normally handled by operators. For maximum productivity, a number of six to ten displays on a substrate is close to optimum, so that 15- to 20-in. displays are most efficiently produced on Generation 5 lines, while Generation 6 and 7 lines are optimized for 20- to 46-in. displays for LCD televisions.

There has not been a standard substrate size in the LCD industry as in the semiconductor industry, where well-defined dimensions for 6-, 8-, and 12-in. wafers have been adopted. As a result, Asian factories use up to 20 different sizes of mother glass substrates. More standardization may occur in the future to reduce the robot, tooling, and equipment cost.

With the newer Generation 5, 6, and 7 lines in production, some of the older production lines have become less efficient for monitor and notebook display manufacturing. These lines are, in many instances, converted to production of smaller 2- to 10-in. displays for cell phones, PDAs, digital cameras, DVD players, camcorders, automobile navigation, entertainment systems, and other consumer electronics products. Some are also upgraded to manufacture LTPS LCDs.

The mass production of AMLCDs has led to the establishment of an infrastructure of suppliers, particularly in Asia. A number of equipment companies provide thin film deposition, patterning, and etching equipment, and LC assembly and module assembly machines. Materials providers include glass and color filter manufacturers, polarizer, retardation and enhancement films vendors, backlight manufacturers, and driver IC and controller IC vendors.

Figure 3.32: Operators in LG Philips LCD Generation 5 factory holding a TFT array substrate with a size of 1050×1200 mm with 15 display patterns of 15-in. diagonal size each.

Their products continue to improve in performance and cost and are essential to the growing success of AMLCD technology.

AMLCD Electronics

The design of the display glass assembly plays an important role in the image quality of an AMLCD. Equally important are the display electronics, which must supply accurate data signals to each pixel. The incoming video signals are either computer-generated or processed from an image acquisition by, for example, a digital camera, television camera, or camcorder. The task of the display electronics is to obtain optimum image quality based on the incoming video signal information.

4.1 Drive Methods

As mentioned earlier, each pixel in the AMLCD is driven with a square wave AC voltage. Ideally, the residual DC component on the pixel voltage is negligible. In practical displays it is unfortunately almost impossible to eliminate the DC component entirely for all gray levels and across the entire display area. The result is that the transmittance of the pixel varies slightly between odd and even frames, as shown in Fig. 4.1.

When the refresh rate is 60 Hz, the frequency of this luminance variation will be 30 Hz since the complete cycle of positive and negative charging of the pixel occurs at 30 Hz. The human eye perceives luminance variations of a few percent at less than 40–50 Hz as flicker. The luminance variations can be smoothed out by using an LC fluid and cell structure with a slow response time, so that the luminance amplitude variations in Fig. 4.1 are below the flicker limit. However, this would cause severe smearing of fast-moving video images and defeats one of the purposes of active matrix drive methods.

Another problem with the DC component on the pixels is that it can cause image sticking or image retention. This leads to burn-in effects on the LCD when a still image is displayed for a prolonged period of time.

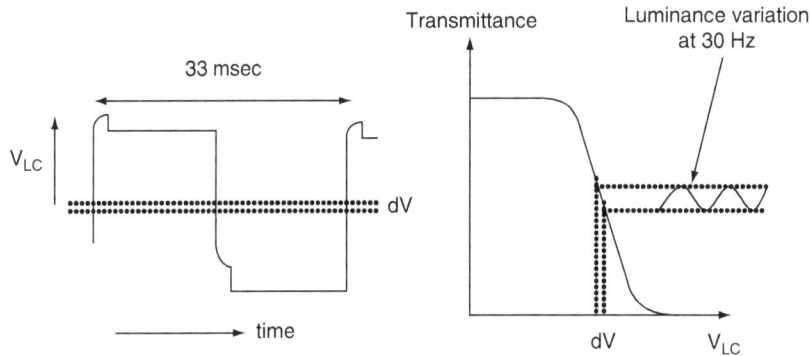

Figure 4.1: Effect of DC voltage component dV of LC voltage on luminance variation.

One way to eliminate flicker is to operate the display at 90 Hz or higher frame rates. Then, the luminance variation will occur at 45 Hz or higher frequencies and will not cause flicker. Since most video and graphics information is supplied to LCD panels at lower refresh rates, this is often not a practical solution.

The most common method to eliminate flicker is the use of inversion drive methods. In this approach, data voltages are applied which alternate the polarity of the voltage on adjacent pixels, so that the human eye perceives an area of multiple pixels as flickerless as a result of spatial averaging. With these polarity inversion methods, individual pixels can still have flicker when viewed close up. At normal viewing distances the flicker becomes imperceptible.

The flicker is minimized or eliminated with this method but image sticking is not reduced; thus, minimizing the DC component on the pixel voltage to less than about 50–100 mV remains crucial.

The polarity can be alternated per row (row or line inversion), per column (column inversion), or by a combination of both (pixel or dot inversion) (Fig. 4.2). Dot inversion gives the best image quality in terms of flicker and also minimizes both vertical and horizontal cross-talk on the display.

In color displays it is important that subpixels of one color have alternating polarity for adjacent pixels, so that flat fields of one color will not exhibit flicker. In a vertical stripe configuration with red (R), green (G), and blue (B) subpixels, this condition is automatically satisfied with dot inversion and also with full pixel inversion, also called color group inversion (Fig. 4.3). As can be seen, in each frame each of the primary color subpixels has opposite polarity compared to the nearest neighbor subpixel with the same color.

The timing and polarity of the data signal on odd and even data buslines is shown in Fig. 4.4 for the different inversion methods, in the case that all pixels are driven at the same gray level. In frame inversion, all pixels receive the same polarity data signal for odd frames and the opposite polarity for even frames, resulting in flicker. In column inversion, adjacent columns have opposite data polarity and change polarity for each frame. In line or row inversion, the data polarity on each pixel changes every line in addition to every frame. Dot inversion combines row and column inversion. In all cases, the driving wave forms on the select lines (the rows) are the same.

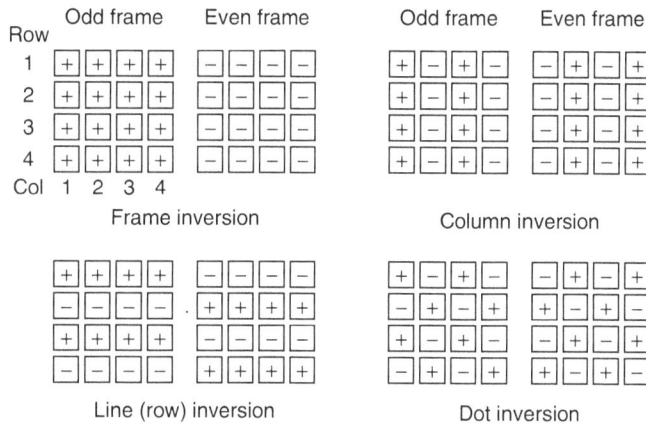

Figure 4.2: Various inversion drive schemes.

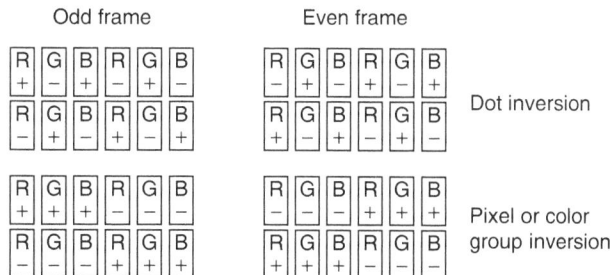

Figure 4.3: Dot inversion (left) and pixel (color group) inversion (right) in color LCDs with an RGB vertical stripe color filter arrangement.

89

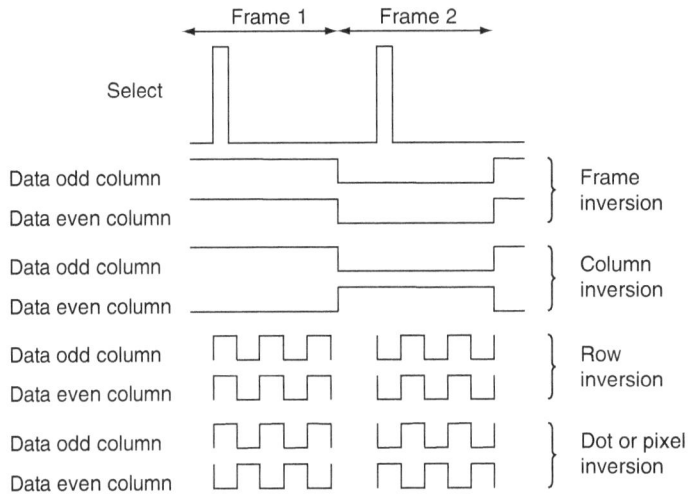

Figure 4.4: Timing of signals in various inversion drive schemes when all pixels are switched on.

In panels for desktop monitors, LCD televisions, and most notebook displays, the voltage on the common ITO electrode on the color plate is held at a constant DC bias voltage V_{com} (Fig. 4.5). The entire voltage range of more than 10 V between the extremes of the positive cycle and the negative cycle must be supplied by the data driver. In actual display operation, the V_{com} voltage is offset by about 1 V in a-Si TFT LCDs to compensate for the pixel voltage shift described in Chapter 2, Sec. 2.6.

In order to reduce driver cost and keep data driver output voltages low, some notebook panels and most small mobile applications employ V_{com} modulation (also called common plane switching), in which the voltage on the color plate switches each line time by about 5 V. Part of the voltage V_{LC} across the LC is then derived from the V_{com} modulation:

$$V_{LC}(+) = V_{data}(+) - V_{com}(+), \qquad (4.1)$$

$$V_{LC}(-) = V_{data}(-) - V_{com}(-). \qquad (4.2)$$

Since the voltage across the LC pixel is the difference between V_{com} and the data driver signal voltage, this allows the use of low-cost, low-voltage data drivers with a full range of 5 V or even 3.3 V when the LC cell has a low operating voltage. A driver IC that can be made in a standard 3.3-V process can be easily procured from many IC fabs at low cost. When the voltage on the color plate is changed every line time, the polarity of the data

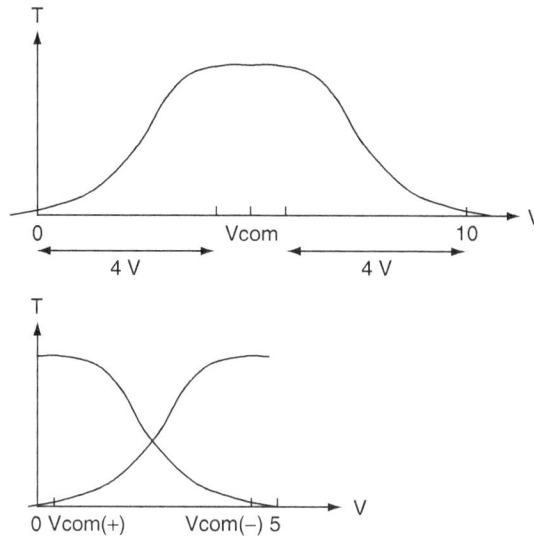

Figure 4.5: Transmittance versus data voltage for full range drive (top) and V_{com} modulation drive (bottom).

voltages supplied to the pixels needs to change every line time as well. V_{com} modulation at the line frequency is therefore used in conjunction with line inversion (row inversion) data drivers. The timing for V_{com} modulation is shown in Fig. 4.6. For activated (ON) pixels, the data voltage is out of phase with the V_{com} modulation, while for non-activated (OFF) pixels it is in phase.

A drawback of V_{com} modulation is that it is incompatible with dot inversion and requires extra electronics to switch the large capacitance load of the V_{com} electrode with the correct timing. This would become a problem for large displays. Most notebooks and all desktop monitors and large LCD televisions therefore use a constant V_{com} and dot inversion (also because dot inversion generally gives the best image quality).

The V_{com} bias voltage is set to compensate for the pixel voltage shift when the select pulses are turned off so as to minimize the residual DC component on the LC pixel voltage. The pixel voltage shift depends on the data voltage, as outlined in Chapter 2, Sec. 2.6 and is different for negative and positive voltage across the LC pixel. To minimize DC components for all gray levels at both pixel polarities, the gamma reference voltages that control gray scale are often set independently for each data polarity. The gamma reference voltages match the incoming video data to the

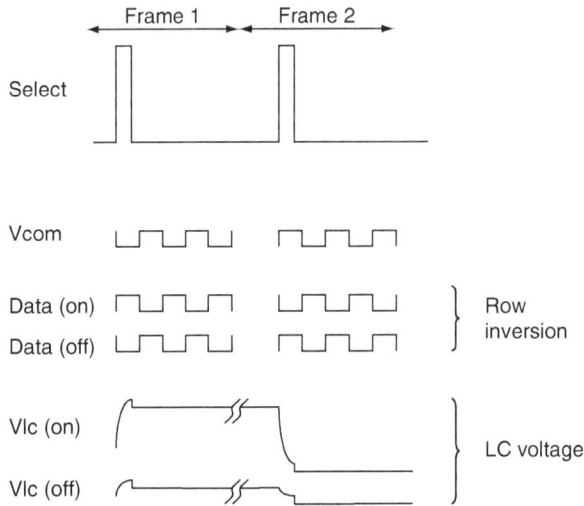

Figure 4.6: Timing of signals and LC voltage in Vcom modulation schemes for all pixels on and all pixels off.

desired gray scale. There are usually about 10 gamma reference voltages. For an 8-bit, 256-level gray scale, the intermediate gray scales are obtained from interpolation between the gamma correction voltages.

This is illustrated in Fig. 4.7, where V_1, V_2,...,V_{10} are the gamma reference voltages, V_{ref} is the video center reference voltage, $AVdd$ is the supply voltage, and $AVss$ is ground. Best performance is achieved when V_1, V_2,...,V_{10} are independently optimized, which means that, in general, $|V_{ref} - V_6| \neq |V_{ref} - V_5|$, $|V_{ref} - V_7| \neq |V_{ref} - V_4|$, etc.

In order to get more than eight bits of gray scale, a technique called frame rate control (FRC) has been developed. With FRC, the gray level voltage can be different between odd and even frames or between four subsequent frames, so that intermediate gray levels can be generated. For example, if during the odd frame the voltage for gray level i and during the even frame the voltage for gray level i+1 is applied, an extra gray level between levels i and i+1 is effectively displayed. This makes it possible to have a 9-bit gray scale with an 8-bit data driver.

The difference in voltages between levels i and i+1 is generally too small to cause a significant DC component on the LC voltage or to cause flicker. The eye will perceive the intermediate gray level.

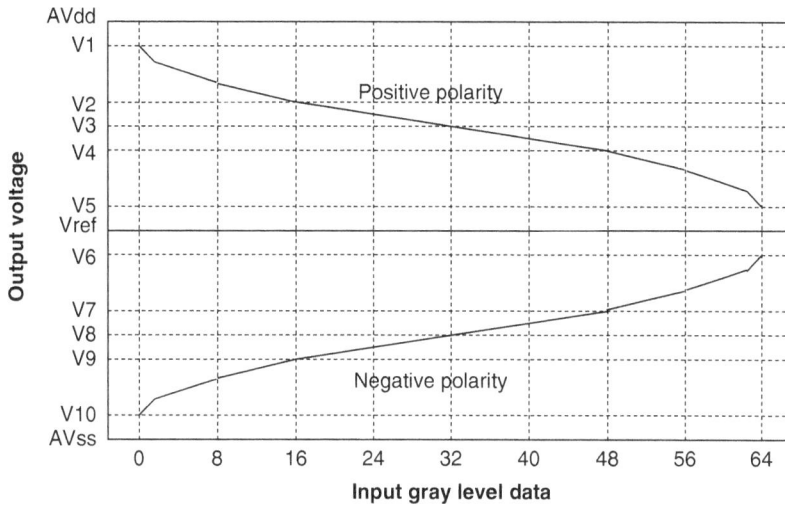

Figure 4.7: Gamma correction diagram.

Similarly, this technique can be used during four subsequent frames to generate a 10-bit gray scale with an 8-bit data driver.

4.2 Row Select and Column Data Drivers

In Fig. 4.8 a block diagram of a TFT LCD module is shown. Directly attached to the display glass are the driver ICs for the rows and the columns. The row driver IC generates the select pulses sequentially addressing each row in the display. It is often referred to as the gate driver, since it switches the TFTs ON and OFF by applying a select voltage to the gate; in the case of a-Si TFTs, this is a positive pulse of about 20 V.

The column drivers supply the video data signals to the data lines connected to the source of the TFTs at each pixel. They are commonly referred to as the source drivers. Gate and source drivers are mixed-signal ICs designed typically with 0.25–1 μm design rules. For most driver ICs there is no need to use advanced design rules below 0.25 μm. The die area is pad limited (i.e., the large number of outputs and their spacing govern how much silicon real estate is required, not the relatively simple electronic circuitry on the chip).

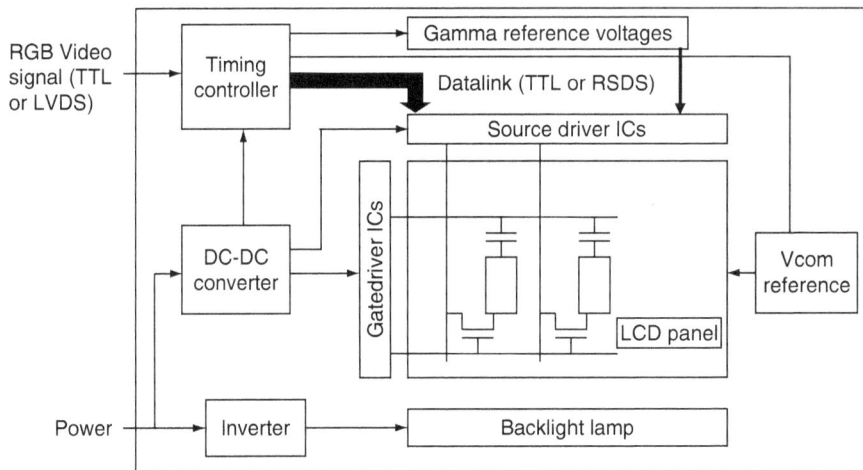

Figure 4.8: Block diagram for a TFT LCD module.

An XGA TFT LCD with 768 rows and 1024×3 columns typically requires three gate driver ICs with 256 output channels each, and eight source driver ICs with 384 channels each. They can be packaged in a tape carrier package (TCP) for attachment to the glass with tape automated bonding (TAB). Alternatively, they are directly bonded upside-down to the glass with chip-on-glass (COG) technology. Since both row and column drivers require a large number of output channels on one side of the chip, the aspect ratio of the the die is usually high (e.g., 20 mm $\times$ 1 mm).

In Fig. 4.9 a block diagram of a gate driver is shown. It has a relatively simple architecture with a bi-directional shift register, level shifters, and output buffers. The shift register transfers a start bit at STVD (which can be considered a vertical sync pulse) through the chip, selecting one row line at a time. When the select bit emerges at the end of the chip it is transferred through STVU to the second row driver IC, where it functions as the start bit. Multiple chips can be cascaded in this fashion.

In this cascade type of operation, all 768 rows in the XGA display are sequentially addressed. The direction of the start bit can be reversed by the DIR input, so that the vertical mirror image can be displayed on the LCD. The DIR function also ensures that the row driver ICs can be mounted either at the left or the right side of the display area.

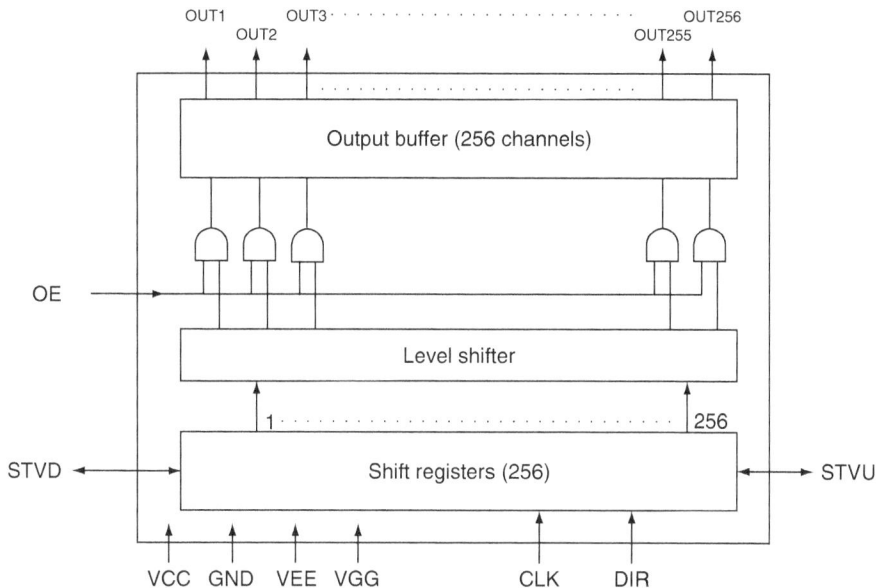

Figure 4.9: Basic architecture of a gate driver IC with 256 channel outputs for use in XGA and SXGA panels (three or four chips per panel, respectively).

The OE input to the row driver allows the select pulse for the rows to be turned OFF before the end of the line time. This is especially important for large displays where the RC delays on the rows require the rows to be turned OFF for a few microseconds before the data voltage on the columns is changed in order to avoid cross-talk. The level shifters raise the logical voltage levels of 0 and 3.3 V to output levels of −5 V and about 20–25 V, respectively. The final levels are compatible with the required ON and OFF voltage levels on the gates of the a-Si TFTs. Output buffers reduce the output impedance to a sufficiently low level in order to drive the gate lines on the AMLCD.

The clock frequency of the row select driver is quite low (around 50 kHz), not much higher than the product of refresh rate (e.g., 60 Hz) and the number of rows (768 for an XGA display). In Table 4.1 the input and output pins of a typical gate driver are listed. A timing diagram is shown in Fig. 4.10.

The digital data driver for the columns, connected to the sources of the TFTs, is somewhat more complex. Its basic block diagram is shown in Fig. 4.11. It basically converts serial digital video input signals into parallel analog data signals for one row. These analog signals representing the intended gray levels are then all simultaneously

Table 4.1: **I/O pins of a basic row (gate) driver IC for a-Si TFT LCDs**

Pin	Input/ Output	Description	Typical value
CLK	I	Clock signal controlling shift register—equal to line frequency	50 kHz
DIR		Direction of operation of bi-directional shift register	0 or VCC
STVD	I/O	Input start bit when operating when DIR=1, output bit for starting next cascaded row driver IC when DIR=0	0 or VCC
STVU	I/O	Input start bit when operating when DIR=0, output bit for starting next cascaded row driver IC when DIR=1	0 or VCC
OE	I	Output Enable to enable output voltage to the TFT LCD rows	0 or VCC
OUT1,..., ...,256	O	Output channels driving the rows (gates) of the TFT LCD	VGG or VEE
VCC	I	Logical voltage supply	3.3 V
GND	I	Ground voltage for logical supply	0 V
VEE	I	Low analog supply voltage for rows	−5 V
VGG	I	High analog supply voltage for rows	20 V

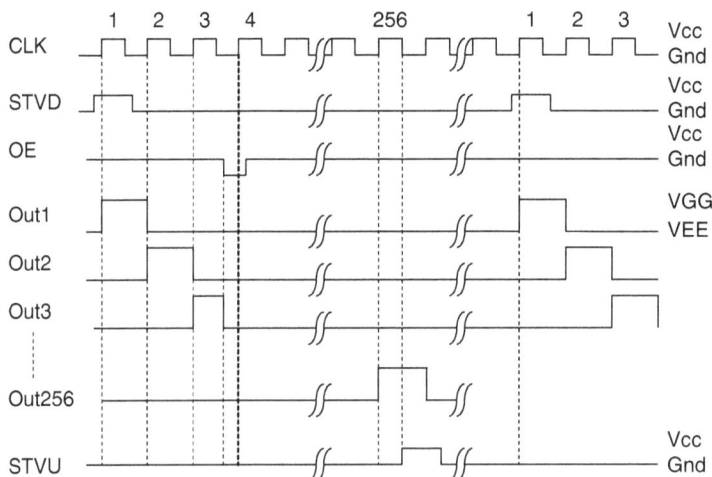

Figure 4.10: Timing diagram of a row driver.

96

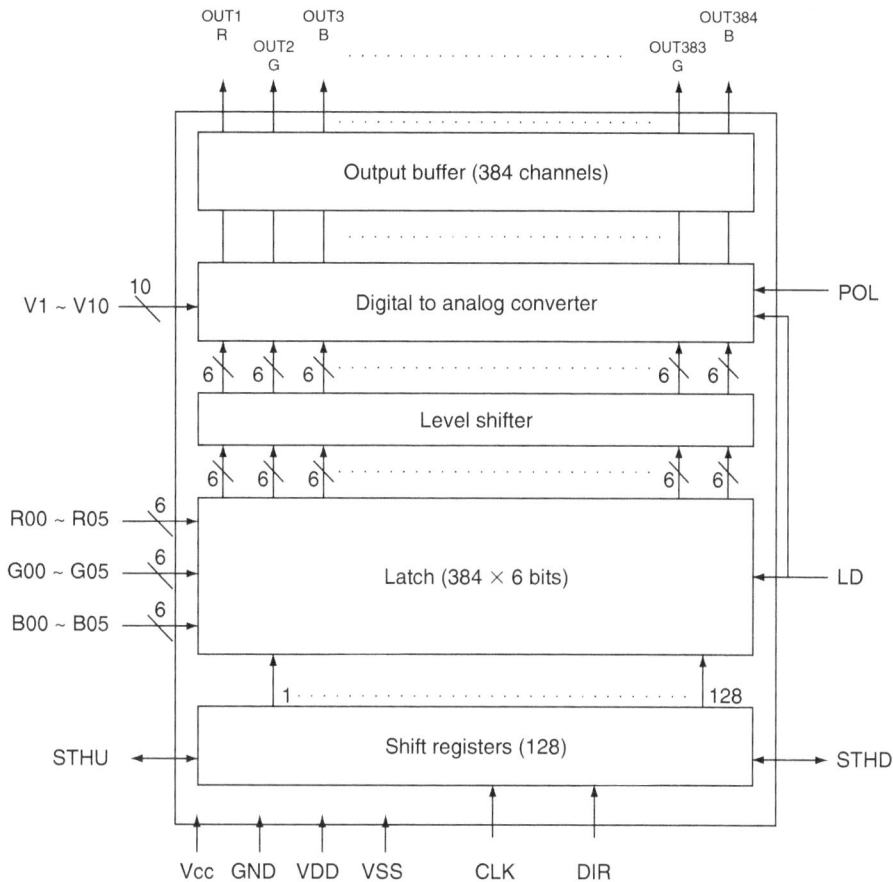

Figure 4.11: Basic architecture of a source driver IC with 384 channel outputs for use in XGA and SXGA panels (eight or ten chips per panel, respectively).

applied to one row of pixels when the row is selected. Like the gate driver, the source driver has a bi-directional shift register with cascading functionality to connect multiple chips together and allow mounting at either the top or the bottom of the viewing area. The 6-bit digital data signals for red, green, and blue channels are latched into a 128×3-line latch and then level shifted from the logical levels to the VDD level. The D/A converter has input reference voltages to convert the digital levels to the appropriate gray scale levels.

The reference voltage inputs to the D/A converter (DAC) ensure the desired gamma correction. The POL input to the DAC controls the polarity of the output signals so that dot, line inversion, or column inversion may be selected. Output buffers reduce the output impedance of the output channels in order to rapidly charge the column line capacitance of the LCD to within a fraction of one line select time.

The operating frequency of the source driver is much higher than for the row driver (around 65 MHz), larger than the product of refresh rate (e.g., 60 Hz) and the number of pixels (768×1024 for an XGA display).

In Table 4.2 the input and output pins of a typical source driver are listed. A timing diagram is shown in Fig. 4.12.

For notebook displays, 6-bit data drivers with 6-bit gray scale (18-bit colors) are common, while for desktop monitors and LCD televisions, 8-bit data drivers (24-bit colors) are used.

Table 4.2: I/O pins of a basic 6-bit column (source) driver IC for a-Si TFT LCDs

Pin	Input/ Output	Description	Typical value
CLK	I	Clock signal controlling shift register—equal to pixel dot frequency	65 MHz
DIR	I	Direction of operation of bi-directional shift register	0 or VCC
STHD	I/O	Input Start bit when operating when DIR=1, output bit for starting next cascaded row driver IC when DIR=0	0 or VCC
STHU	I/O	Input Start bit when operating when DIR=0, output bit for starting next cascaded row driver IC when DIR=1	0 or VCC
POL	I	Selects polarity of the output channel voltages to the columns	0 or VCC
LD	I	Switches new data signals to the outputs and latches their polarity	0 or VCC
V1, ... V10	I	Voltage bias levels to set gamma correction	
R00–R05, G00–G05, B00–B05	I	Digital video input signals for red, green, and blue color pixels	
OUT1,...., ...,384	O	Output channels driving the columns (sources) of the TFT LCD. OUT1,4,7,... for red pixels, OUT2,5,7,... for green pixels, OUT3,6,9,... for blue pixels	VDD<OUTx< VSS
VCC	I	Logical voltage supply	3.3 V
GND	I	Ground voltage for logical supply	0 V
VSS	I	Low analog supply voltage	0 V
VDD	I	High analog supply voltage	10 V

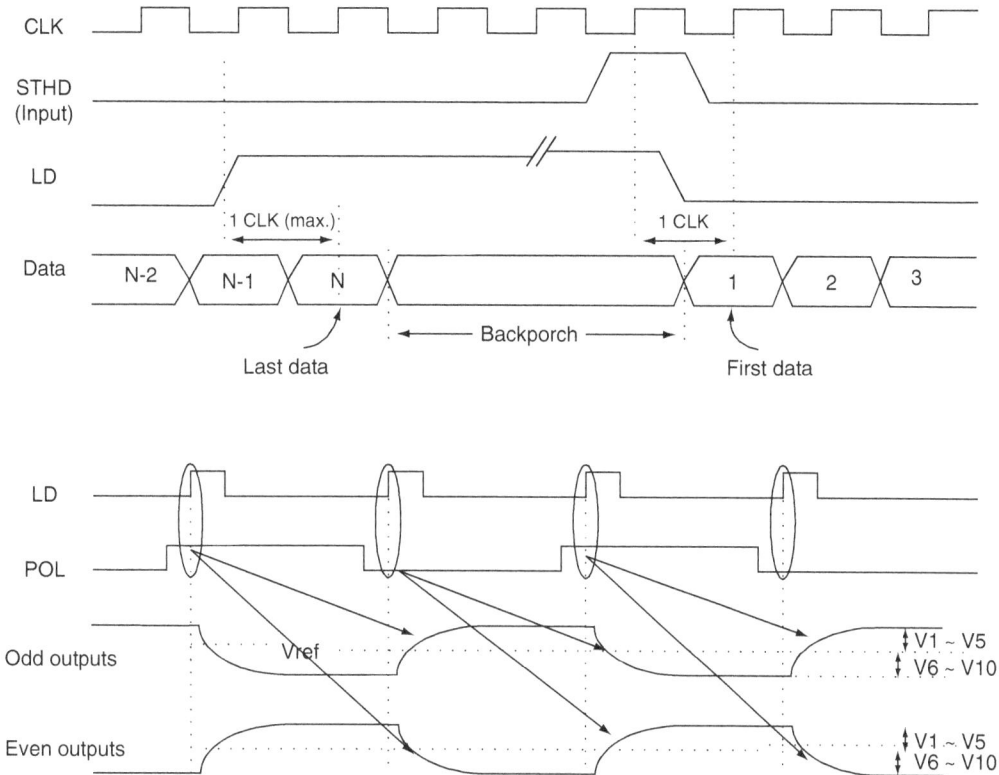

Figure 4.12: Example of a timing diagram for a column driver. Top four wave forms show clocking in of data. Bottom four wave forms show latching of analog output voltage to the output using the POL signal to obtain dot inversion.

Analog data drivers, which receive an analog video signal, are used in some consumer electronic LCDs, but are gradually being phased out.

4.3   Timing Controllers, Display Controllers, and Interfaces

In the block diagram of Fig. 4.8, the timing controller (TCON) is the chip that supplies the control and video signals to the row and column drivers. For the row driver, they are the vertical clock, the vertical start bit, the output enable signal, and the logical bit that controls the direction of the shift register (see Table 4.1). For the column driver, they include the horizontal clock, the input start bit, the polarity signal to control the polarity of the data voltage applied to the LCD, the gamma reference control voltages, and the digital video data signals (see Table 4.2).

For low-resolution displays, the data link between TCON and the column driver is in transistor-transistor-logic (TTL) format, which switches the digital signal typically between 0 and 3.3 or 5 V. For higher-resolution displays beyond SVGA (600×800 pixels), the frequency of the data signals becomes so high for TTL as to cause electromagnetic interference (EMI) and signal fidelity problems. Therefore, XGA and higher-resolution displays use a method of differential signaling to cancel out most of the EMI. These interface architectures are based on low voltage differential signaling (LVDS) [1] or reduced swing differential signaling (RSDS) standards [2]. RSDS has become the de facto standard in most cases for the video interface between the timing controller and the data drivers.

The timing controller receives a video signal which is already compatible with the resolution of the LCD panel. Since notebook displays are permanently attached to the laptop computer, these digital signals are directly generated by the graphics controller circuitry in the computer.

Figure 4.13 shows a photograph of an XGA TFT LCD module or panel, as supplied by the AMLCD manufacturer to notebook manufacturers. It includes the backlight unit but not the inverter for the backlight. The backlight consists of a cold cathode fluorescence (CCFL) stick lamp at the edge of a light guide, which evenly distributes the light across the viewing area of the panel. This particular module has only one PCB and uses chip-on-glass row and column drivers. Flexible interconnect tape connects the PCB to the inputs of the data drivers and scan drivers on the glass. The control signals for the

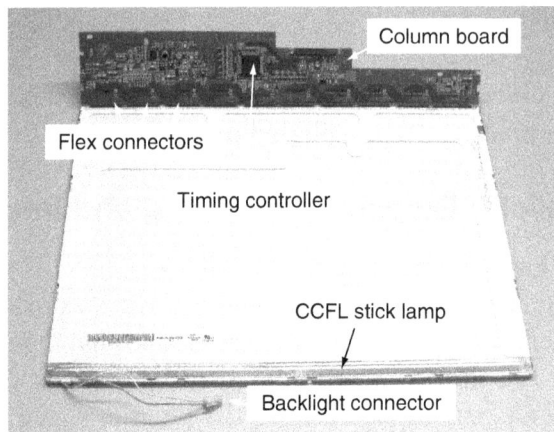

Figure 4.13: View of back of a TFT LCD module for notebook displays.

three scan drivers are routed on the glass along the right side of the panel. The PCB is folded back in the photograph to expose the timing controller. Normally, the PCB is folded flat against the panel to obtain a very thin module with a thickness of only a few millimeters.

For standalone LCD desktop monitors, the design is somewhat different. They need to be compatible with hookup to a variety of computers with analog and digital video output formats and resolutions. Usually the module, without the inverter, is supplied to display integrators who build the complete monitors. They add the inverter to operate the backlight, a monitor controller board, and a front panel control board. The unit is then mounted in the final enclosure with a stand. The front panel control has the ON/OFF switch, brightness, contrast, and hue controls.

Figure 4.14 shows the back of a display panel for desktop monitors, including the backlight inverter, front panel control board, and the monitor controller board. The backlight inverter drives stick lamps on both sides of the panel to obtain sufficient brightness. The monitor controller board supplies processed video signals to the timing controller on the column board.

The monitor controller board accepts different video signal formats and converts them to the proper format for the timing controller. For hookup to computers, these formats include at least analog RGB video (also called analog VGA) and, for dual mode

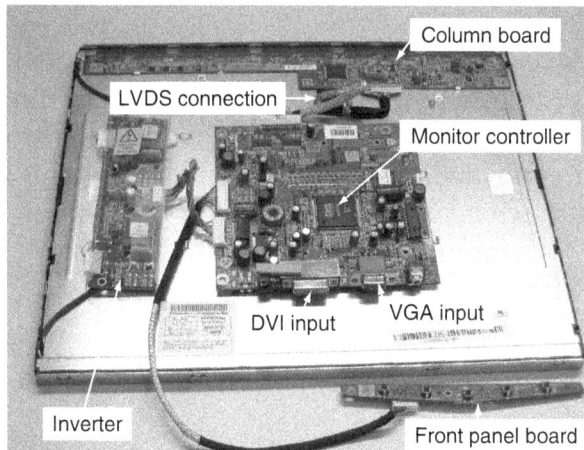

Figure 4.14: View of back of a TFT LCD panel for a desktop monitor including monitor controller board, inverter, and front panel control board.

monitors, they also include digital video interface (DVI) [3]. The analog VGA and DVI connectors are indicated in the photograph. In some cases there also inputs for S-video, component video, and composite video for operation as a television and for connection to external DVD players and other consumer electronics.

The main chip on the monitor controller board usually can perform scaling functions, convert the different formats, provide on-screen display (OSD), picture-in-picture (PIP), and other functionality. It has a built-in microcontroller and can perform some image processing, color management, and gamma control functions. In some cases, LVDS transmitters are built in. The interface format for the data-link between the monitor controller board and the timing controller is LVDS to minimize EMI.

Detailed descriptions of video and interface formats are outside the scope of this book; for more information, the reader is referred to the literature and several excellent books [4,5]. It should be noted that new, improved standards in video data transfer techniques and interfaces are still periodically introduced. For example, for notebook panels and LCD television, the point-to-point differential signaling (PPDS) interface architecture has been proposed by National Semiconductor [6] to reduce the number of inputs to each data driver by more than 60%. This is achieved by a point-to-point distribution of the digital video signals. It reduces the number of video inputs to each chip from 18 to 2 for 6-bit video. In addition, the gamma reference voltages are digitally generated in the timing controller rather than hardwiring them as 10 analog voltages to each data driver. D/A converters inside the data driver generate the analog gamma reference voltages from one digital serial input. For the PPDS interface, new timing controllers and data driver ICs have been developed.

In the LCD monitor market there is an alternative business model in which more of the monitor integration is performed at the AMLCD panel manufacturer. In this so-called smart panel approach, a single display controller chip is used that combines the functions of the monitor controller chip and the timing controller chip. This leads to a higher level of integration at the expense of design flexibility.

4.4   Integration of Electronics on Glass

As mentioned in previous chapters, some of the peripheral electronics of the TFT LCD can be integrated on the glass. The low mobility of a-Si TFTs precludes the design of high-speed circuitry with a-Si TFTs.

Despite this hurdle, there have been successful efforts to integrate a-Si TFT row drivers, and this technology has now been commercialized in small displays and notebook panels

[7]. The row driver circuits require only 10–30 kHz clock frequency, which can be easily achieved with a-Si TFTs. The major challenge with a-Si TFTs is to avoid large threshold voltage shifts in TFTs with high duty ratios in shift registers, buffers, and level shifters. They reduce the lifetime of the display. A careful consideration of the a-Si TFT threshold shift and an optimized row driver circuit is needed to get a lifetime exceeding 10,000 hours [8]. There have also been reports on displays in which both row and column drivers are integrated with a-Si TFTs [9]. By integrating peripheral electronics on the glass, the reliability of the display system is enhanced since there are fewer interconnects and interconnect failures are less likely.

The vast majority of integration efforts have focused on poly-Si TFTs, mostly of the low-temperature variety—low-temperature poly silicon (LTPS). In Fig. 4.15 the operating frequency of a shift register is plotted as a function of the poly-Si TFT mobility and for different design rules of the TFT channel length L.

As explained in Sec. 4.2, the operating frequency of the row driver circuits does not exceed more than about 50 kHz, which is easy to achieve with LTPS TFTs and even with a-Si TFTs. The data drivers are more difficult to implement, but have been successfully integrated on the glass in small mobile commercial products, such as PDAs, with a relatively low pixel count. In Sec. 4.2 it was shown that the data driver for an XGA display requires a clock frequency of 65 MHz. According to Fig. 4.15, it would take a

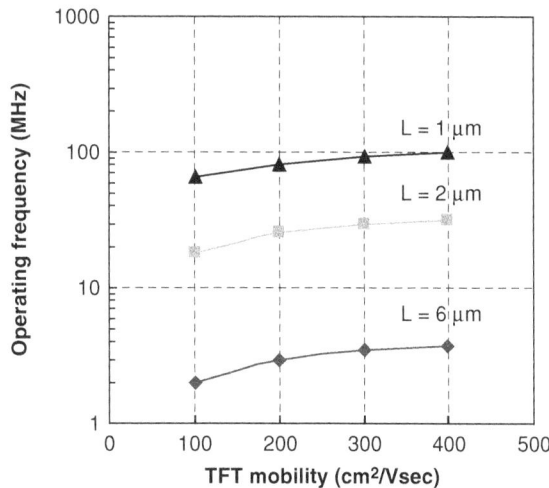

Figure 4.15: Dependence of shift register operating frequency on TFT mobility for different design rules of the gate length L

process with 1-μm design rules to integrate the data drivers in an XGA display and a TFT mobility more than 100 cm^2/Vsec.

Increasingly, other functionality such as D/A converters, timing controllers, graphics controllers, and DC/DC converters are also considered for integration. Displays with a higher level of integration are often referred to as "system-on-panel" or "system-on-glass." Figure 4.16 shows an example of a configuration for a system-on-panel. By adding the DC/DC converter on the glass, a single power supply voltage of, for example, 3.3 V for the entire system is possible. One joint program by Sharp and Semiconductor Energy Labs succeeded in integrating a microprocessor and audio circuitry on the glass [10] as well. Since the design rules for LTPS (1–2 μm) are much less advanced than for state-of-the-art integrated circuits (~0.1 μm), the peripheral circuitry occupied an area larger than the display area and the microprocessor had the speed and performance of a 1980s-era processor. Such a high level of integration is therefore more a capability demonstration than a practical solution.

The decision to include more circuitry on the glass depends on a trade-off between cost and performance. A reduction in manufacturing yield is also possible with more integrated circuitry on the glass and is taken into account.

It should be noted that the functionality achieved with a system-on-panel LTPS display can often be obtained at lower cost using an a-Si TFT LCD with discrete external ICs attached with chip-on-glass bonding. The latter approach is a trend for mobile phone displays where a single chip, with a single power supply and bonded by COG, is used to address the rows and columns in 176(×3)×220 pixel color a-Si TFT LCDs. Single-chip driving is expected to be extended to displays with even more pixels. The single chip

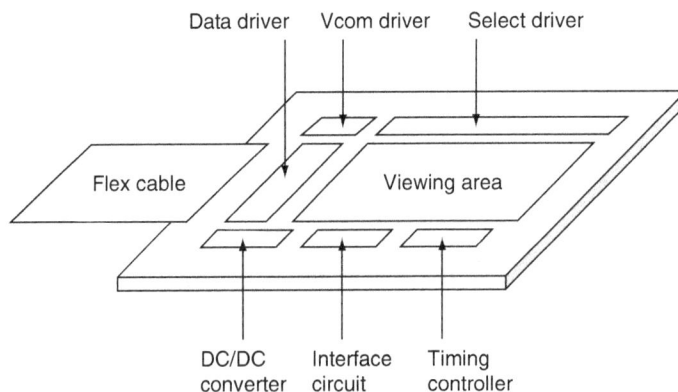

Figure 4.16: Example of configuration for a "system-on-panel."

includes built-in timing controllers, graphics controllers, and other functionality. This trend is supported by IC foundries, which now offer 40-V processes with 0.18-μm design rules so that the power management can also be integrated.

To counter the move toward small, portable applications using a-Si TFT LCDs with single-chip driving, LTPS manufacturers are proposing to include functionality in the viewing area of the display, such as memory built into the pixel, saving power and eliminating the need for continuous refresh of static images. Other approaches include the addition of photosensor arrays in the display viewing area to function as ambient light sensors, scanners, optical touch panels, or fingerprint sensors.

4.5 Backlights

Most AMLCDs operate in the transmissive mode. They can be considered electronically controlled transparencies on a light box—the backlight. It is obvious that many display parameters, including brightness, depend strongly on the backlight. For color displays, the combination of backlight spectrum and color filter spectra determines the color coordinates in the display, as will be outlined in Chapter 5.

In this section we are concerned about the electrical, electronic, and mechanical issues regarding backlights. There are basically two mechanical implementations of a backlight: edge lighting and surface lighting.

An essential requirement for portable, battery-powered applications is to keep power consumption in check and minimize overall panel thickness. This is achieved by edge lighting with light guides. Edge lighting uses one or more lamps, typically cold cathode fluorescent lamps (CCFL) tubes, at the edge of the display, in combination with a thin light guide that distributes the light evenly across the display surface. They allow a thin and lightweight design. Edge lighting with one stick lamp is the backlight of choice for notebook LCDs. They give a display luminance as high as 250 cd/m^2 when used in combination with brightness enhancement films (see also Chapter 6). Stick lamps with a diameter less than 2 mm are available for notebook LCDs.

Figure 4.17 shows the configuration of the edge light with a CCFL stick lamp. The light guide has a diffuse reflector on its back side and a diffuser on the LCD side. For larger displays (exceeding 15 in.) used in monitors, one to three stick lamps on both long edges of the display are often employed in combination with a thicker and heavier light guide.

The other type of CCFL backlight has a cavity configuration with multiple stick lamps or U-shaped lamps (Fig. 4.18) behind the LCD panel. This configuration is used in LCD televisions and in some monitors requiring higher brightness.

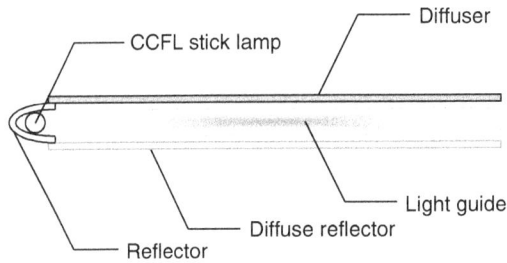

Figure 4.17: CCFL backlight configuration for notebook LCDs.

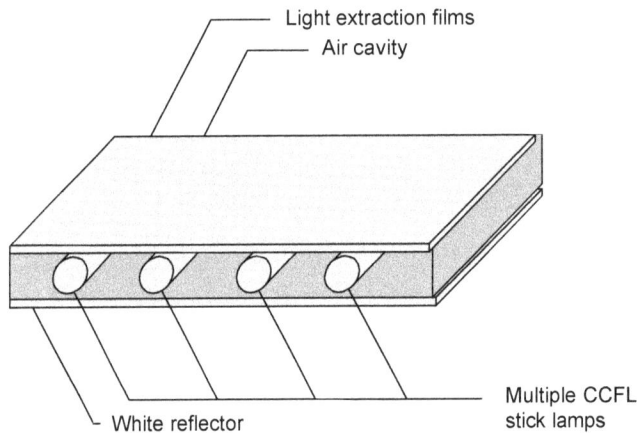

Figure 4.18: Cavity backlight construction.

To obtain uniform luminance across the display area, the light from the lamps needs to be redistributed with a diffuser, leading to some loss in efficiency. In this configuration it is also important that the lamps age approximately the same over time to avoid bands with varying brightness on the LCD after prolonged operation. Large modules with surface backlighting tend to be much lower in weight than large edge-lit panels because the thick, heavy light guide is replaced with an air cavity with stick lamps.

Cold cathode fluorescent lamps (CCFLs) are sealed glass tubes with an electrode at each end. They are filled with an inert gas mixed with a small amount of mercury. To operate the lamp, a high voltage is applied between the electrodes, causing the gas to ionize and emit ultraviolet light. The inner surface of the tube is coated with phosphor layers. The UV rays excite these phosphor layers so that they emit light in their characteristic wavelengths. For color LCDs the CCFL phosphors have three dominant wavelengths corresponding to the primary colors, red, green, and blue. By adjusting the type and mixing ratio of the three phosphors, there is control over the color performance of the backlight. The lamps are made with different diameters and can be linear or have various shapes such as a U-shape or a serpentine shape.

A high-frequency, high-voltage driving wave form to sustain the discharge in the lamp is supplied from a DC/AC inverter, a transformer that converts the DC input voltage of 3.3 or 5 V into a sine wave of more than 1000-V peak-to-peak amplitude. The optimum frequency is around 50 kHz.

Display brightness can be controlled by dimming the backlight. Dimming is achieved by pulse width modulation, which affects the average current through the lamp. Dimming ratios with CCFLs are typically less than 100:1.

CCFLs have an efficiency of around 25%, which corresponds with about 120 lm/W. (See Chapter 5 for a definition of lumens.) The rest of the input power is lost in the discharge and in heating. The operating temperature range is 20 to 60°C. Between the lamps and the display glass, several plastic films are inserted to redirect the light. Some of them diffuse the light to improve uniformity and others enhance brightness in the direction normal to the display at the expense of oblique directions. These films are discussed in more detail in Chapter 6.

Although CCFLs are inexpensive and bright, they have some drawbacks in terms of lifetime, temperature operating range, and the presence of small amounts of mercury, which pose a hazard at disposal.

Several other types of light sources for LCDs have therefore been developed and have entered the marketplace. They include hot cathode fluorescent lamps (HCFLs), mercury-free Xenon-based flat lamps, and LED-based backlights.

HCFL backlights are receiving renewed interest after initial use in high-end applications such as avionic cockpit displays. The main reasons for their use in avionics are that they can have dimming ratios of 10,000:1 and a wider operating range, required for this application.

HCFLs are common in household applications. They include a tungsten filament, which creates thermionic emission of electrons when heated. As a result, they are easier to strike and dim than CCFLs, which rely on secondary emission of electrons. The renewed interest in HCFLs is caused by the move to scanning backlights for LCD televisions to improve motion portrayal (see also Chapter 6, Sec. 6.6.1.). They have typically wider diameters of about 16 mm and the maximum lamp current for an HCFL is in the range of 50 mA to 1 A, significantly higher than the 10–20 mA typical for CCFLs. They can therefore be operated intermittently at higher brightness than CCFLs. The initial drawback of lower lifetimes in HCFLs has been addressed and overcome as well.

The Xenon-based flat lamp [11] has external electrodes and phosphors coated on the inside of flat glass substrates to convert the UV radiation from the Xe plasma into visible light. It has several advantages over CCFLs. While the individual CCF stick lamps in a cavity backlight require diffusers and other measures to improve uniformity, the Xe flat lamp is much more uniform by itself. By using external electrodes, which are not exposed to the plasma, it also has a longer lifetime, approximately 100,000 hours to half brightness, as compared to 50,000 hours or less for CCFLs. The elimination of mercury reduces environmental concerns about disposal. Mercury-free backlights are becoming mandated in a number of countries.

The Xe flat lamp also can provide more saturated colors for a wider color gamut of 75% of NTSC, as compared to 65% for conventional CCFL-based displays (see Chapter 5 for definitions of color performance). The operating temperature range is −20 to 80°C so that they can meet outdoor performance requirements. CCFLs, on the other hand, show a drop-off in efficiency below about 20°C and above 60°C and are therefore less suitable for outdoor applications.

The major drawback of Xe flat lamps is their higher cost, which has prevented their penetration into mass markets. With volume production of LCD televisions, this may change.

A third major backlight technology is based on light-emitting diodes (LEDs). With the dramatic improvement in efficiency of red, green, and especially blue LEDs over the past decade, they have become attractive for many lighting applications, including LCD backlights. LEDs can be operated with low-voltage DC power supplies, and brightness is proportional to the current through the LED. The increasing popularity of LED lighting in general is related to significant improvements in their optical efficiency and packaging, including optimally designed heat sinks which allow much higher currents and brightness. A basic LED emits light in a narrow wavelength region, controlled by the energy gap of the semiconductor material (such GaAlAs or GaInP). White-emitting LEDs can be obtained by a UV-emitting LED in combination with visible light-emitting phosphors.

The challenge for R, G, and B LED backlights is the mixing of the three colors from multiple individual point sources to obtain a uniform backlight. This has been achieved [12], although the power conversion efficiency is still only about 50% of that of a CCFL backlight. Some of the advantages of LED backlights are a wide color gamut (100% of NTSC is possible) and a long lifetime. A drawback so far has been a much higher cost, although the manufacturing cost of the LEDs themselves continues to be lowered. Figure 4.19 shows a spider diagram comparing the characteristics of CCFL, Xe, and LED backlights in terms of lifetime, uniformity, maximum size, color gamut, power consumption, and manufacturing cost (with the caveat that especially the latter two keep improving). Larger values on the diagram indicate more favorable characteristics.

The light from LEDs can easily be pulsed with faster-than-microsecond response times by changing the driving supply from DC to pulsed. LED backlights are therefore also of interest in field-sequential LCDs without color filters, in which the red, green, and blue LEDs are turned on at different times to obtain temporal color mixing. This will be discussed in more detail in Chapter 6.

4.6 Power Consumption

The power consumption in a transmissive TFT LCD module can be subdivided into backlight power, power for the scan driver, the data driver, and the control circuit (Fig. 4.20). Typical power consumption in a 10.4-in. display is about 3 W at 150 cd/m^2 brightness and is dominated by backlight power. This underscores the importance of high-aperture designs and other brightness enhancements to limit battery power use in portable applications.

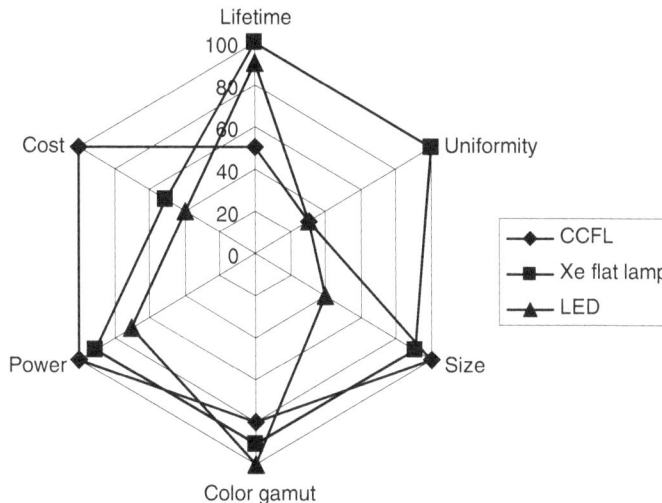

Figure 4.19: Comparison of different backlight technologies.

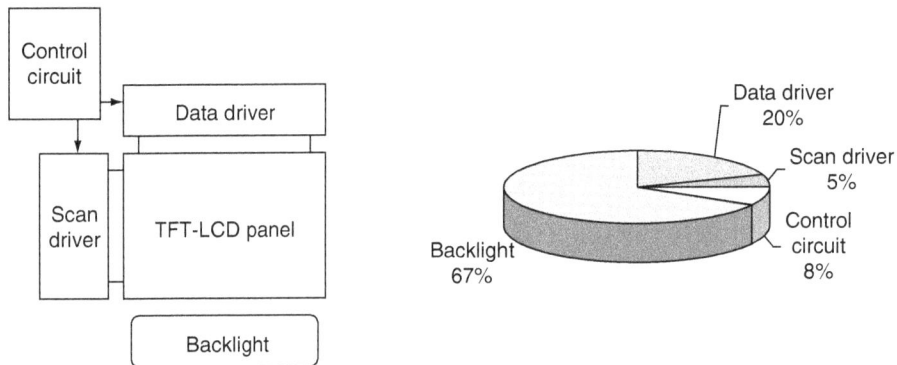

Figure 4.20: Contributions to power consumption of about 3 W in typical 10.4-in. backlit TFT LCD with 150 nits brightness.

A 20-in. LCD television has a power consumption of about 50 W, including all the electronics. This compares favorably with a 20-in. CRT television, which uses about 100 W. It has been calculated that in Japan alone the transition from CRT to LCD televisions will save 5 billion kWh annually.

To further reduce power consumption in portable LCDs, there have been many efforts to eliminate the backlight by developing reflective LCDs with acceptable image quality. They are applied in some handheld products, but the need to view most displays under any ambient lighting conditions has prevented reflective displays from becoming a mainstream technology. A more successful approach is the transflective LCD, in which each pixel is partially transmissive and partially reflective. The reflective area ensures display legibility at high ambient lighting such as outdoor conditions. The transmissive area is used in dark and low ambient lighting conditions. The differences between transmissive, reflective, and transflective displays are discussed in more detail in Chapter 6.

References

1. http://www.siimage.com/documents/SiI-WP-007-A.pdf
2. http://www.national.com/appinfo/displays/
3. http://www.siimage.com/documents/SiI-WP-007-A.pdf
4. K. Jack, *Video Demystified, A Handbook for the Digital Engineer*. Newnes: Elsevier Science (2001).
5. R. Myers, *Display Interfaces: Fundamentals and Standards*. New York: SID-Wiley Digital Video Interfaces (2002).
6. http://www.national.com/appinfo/fpd/ppds.html

7. J. Jeon, K.S. Choo, W.K. Lee, J.H. Song, and H.G. Kim, "Integrated a-Si Gate Driver Circuit for TFT-LCD Panel," *SID 2004 Digest*, pp. 10–13 (2004).

8. H. Lebrun, T. Kretz, J. Magarino, and N. Szydlo, "Design of Integrated Drivers with Amorphous Silicon TFTs for Small Displays," *SID 2005 Digest*, pp. 950–953 (2005).

9. R.G. Stewart, J. Dresner, S. Weisbrod, R.I. Huq, D. Plus, B. Mourey, B. Hepp, and A. Dupont, "Circuit Design for a-Si AMLCDs with Integrated Drivers," *SID 1995 Digest*, pp. 89–92 (1995).

10. T. Ikeda, Y. Shionoiri, T. Atsumi, A. Ishikawa, H. Miyake, Y. Kurokawa, K. Kato, J. Koyama, S. Yamazaki, K. Miyata, T. Matsuo, T. Nagai, Y. Hirayama, Y. Kubota, T. Muramatsu, and M. Katayama, "Full-Functional System Liquid Crystal Display Using CG-Silicon Technology," *SID 2004 Digest*, pp. 860–863 (2004).

11. http://www.osram.de/download/osram_planon_en.ppt

12. http://www.lumileds.com

Performance Characteristics

The image quality and small form factor of TFT LCDs has allowed their application in numerous portable devices as well as in desktop monitors and LCD television. This chapter first describes the basic principles of photometry and colorimetry used to characterize AMLCDs, followed by a description of the basic characteristics of standard AMLCDs in terms of viewing angle behavior, brightness, response time, size, and resolution.

5.1 Basics of Photometry and Colorimetry

Electronic displays generate images in the visible spectrum that are transferred to the eye and the brain. The visible spectrum (or light) is one form of electromagnetic radiation, with wavelengths in the range of 380–780 nm. The intensity of electromagnetic waves or photons can be expressed in radiometric units such as mW/cm^2.

The following explanation of brightness and color in displays is basic and approximate. For a much more thorough and accurate treatment, the reader is referred to one of the many excellent books on this topic [1,2,3].

The human eye is sensitive to only part of the electromagnetic spectrum (the "visible" range), as shown in the eye response (or photopic spectral luminous efficiency) curve $V(\lambda)$ of Fig. 5.1. While radiometry addresses the measurement of radiation in any part of the electromagnetic spectrum, photometry focuses exclusively on the visible light range.

The quantitative measure of display brightness is luminance *Lum*. It is expressed in photometric units:

$$Lum = k \int_{380}^{780} I(\lambda)V(\lambda)d\lambda, \qquad (5.1)$$

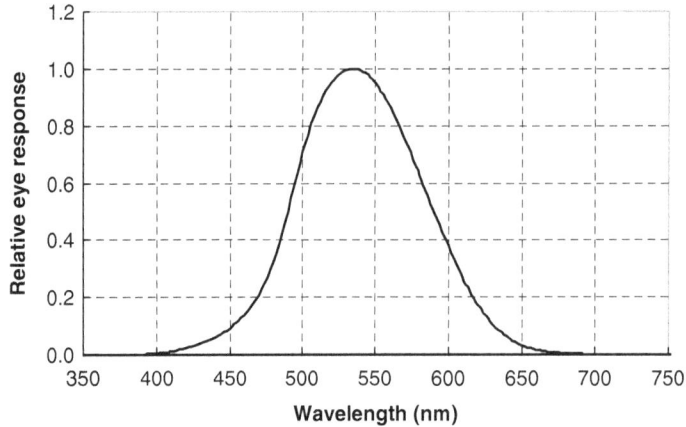

Figure 5.1: Eye response or photopic luminous efficiency curve.

where $I(\lambda)$ is the spectral radiance in radiometric units (Watt/ster.nm.m$^{2.}$), λ is the wavelength, and k is the conversion factor equal to 683 lumens/Watt. The unit of luminance is candela/m^2 (cd/m^2), also called nits. An older unit for luminance, still often used, is footLambert (fL).

Luminance can be considered the brightness emanating from a surface. It should not be confused with illuminance, which is the light intensity incident on a surface. Illuminance is expressed in lux or footCandle (fC). The corresponding terms in radiometry are radiance and irradiance.

Table 5.1 summarizes the basic photometric units. The relation between luminance and illuminance is that 1 footCandle incident on a perfect diffuse (Lambertian) white reflector leads to 1 footLambert of luminance emanating from the surface.

LCDs for notebooks, desktop monitors, and LCD televisions have a luminance of around 200, 350, and 600 cd/m^2, respectively.

Table 5.1: Photometric units

Luminance	Illuminance
1 nit = 1 cd/m2	1 lux = 1 lumen/m^2
1 cd/m^2 = 0.2919 fL	1 lux = 0.0929 fC
1 fL = 3.426 cd/m^2	1 fC = 10.76 lux

The relevance of illuminance in this context is that display contrast depends on ambient lighting conditions because of reflections from the display surface. Average room light conditions correspond to about 500 lux, and bright outdoor lighting to an illuminance of up to 100,000 lux.

Like photometry, the science of measuring color (colorimetry) is based on experiments with human observers having normal vision. It was found from color matching experiments that three independent variables (or tristimulus values) X, Y, and Z are necessary and sufficient to describe color and luminance.

Chromaticity coordinates are derived from X, Y, and Z as follows:

$$x = \frac{X}{X+Y+Z}, \quad y = \frac{Y}{X+Y+Z} \quad and \quad z = \frac{Z}{X+Y+Z}. \tag{5.2}$$

Since $x + y + z = 1$, two chromatic coordinates are sufficient to specify a color. The ratios x and y are used for this. It should be noted that chromaticity coordinates give only ratios of tristimulus values and therefore provide no luminance information.

In 1931 the Commission Internationale de l'Eclairage (the standard body CIE) published the standard CIE 1931 Chromaticity Diagram, based on these concepts (see Fig. 5.2). It is a horseshoe-shaped curve that encompasses all possible color mixtures visible to humans. The purest colors from single monochromatic light are located on the curve itself, with blue in the bottom left, green near the top of the curve, and red at the right side. Closer to the center of the diagram mixed colors, including white, are represented. The triangle shown inside the chromaticity diagram would represent the color performance of a color display. Only colors within the triangle (the color gamut) can be displayed. By turning off two of the three R, G, and B color subpixels, the chromaticity on the extreme points of the triangles (the primary colors) are obtained. Depending on the number of gray levels, the combination of the three primary colors allows presentation of a number of intermediate colors by additive color mixing.

For example, for n-bit gray levels 2^{3n} colors can be portrayed. This implies that on a display with a 6-bit gray scale, 262,144 colors can be displayed, corresponding to 4096 different points in the chromaticity diagram with 64 luminance levels each. An 8-bit gray scale will display 16,777,216 colors, corresponding to 65,536 points with 256 luminance levels each.

Obviously, the larger the triangle (the color gamut), the better the display can show saturated colors closer to the horseshoe-shaped curve.

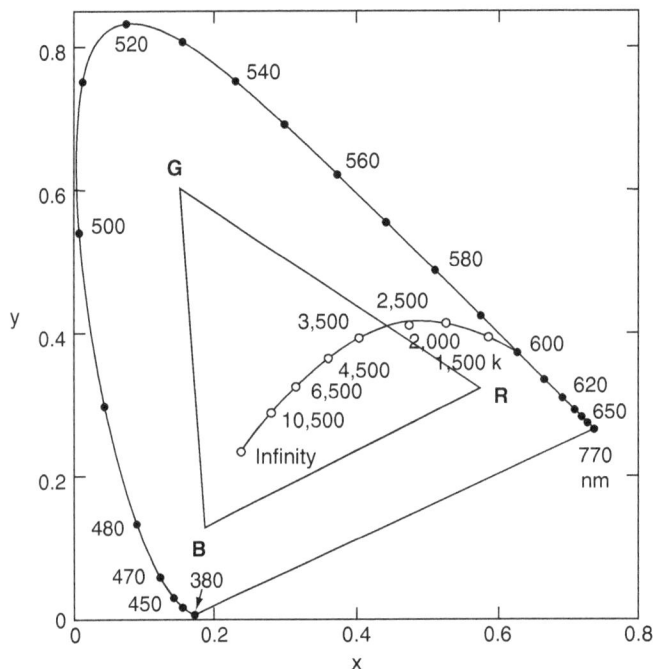

Figure 5.2: CIE 1931 Chromaticity Diagram with blackbody curve and triangle representing primary colors of a color display.

When display contrast ratio is low, two of the three colors cannot be turned OFF completely when the third color needs to be displayed. This will shrink the color triangle as a result of undesired color mixing. The display will then not be able to achieve the color purity possible with the individual color filters.

The chromaticity diagram of Fig. 5.2 also shows the so-called blackbody curve, which is the color of a radiating object at increasing temperature in degrees Kelvin (K). The curve moves from a red glow below 1000 K to a white point at 10,000 K.

A chromaticity coordinate on or near the blackbody curve is referred to as the color temperature and is an important parameter in the specification of the white point of LCD televisions. It should be noted that in LCDs the chromaticity triangle and the color temperature will normally vary with viewing angle, gray level, and ambient lighting. One of the main efforts to improve LCD performance is concerned with minimizing these variations (see Chapter 6).

Each point in the chromaticity diagram can be surrounded by a small ellipse, which represents the border of so-called "just noticeable differences" (JNDs). Any variation of the color coordinates *x,y* within this ellipse will not be noticeable to the average observer.

In the CIE 1931 diagram, the ellipses representing JNDs have different sizes and shapes depending on their location in the chromaticity diagram. In other words, the JNDs are non-uniform.

To obtain uniform JNDs, the alternative CIE 1976 Chromaticity Diagram with *u'*, *v'* coordinates was introduced by the CIE, as shown in Fig. 5.3. The *u',v'* coordinates can be obtained from the *x,y* coordinates with a simple linear transformation [2]. In the *u',v'* diagram, JNDs are represented by a circle with the same radius at any location (i.e., by excursion of a fixed $\Delta u'$, $\Delta v'$ from a particular coordinate).

5.2 Brightness and Contrast Ratio

Since most TFT LCDs are backlit, their brightness is proportional to the backlight intensity. Peak luminance of 150–400 cd/m^2 is typical for notebook and monitor displays (Fig. 5.4). The white luminance is the sum of the red, green, and blue luminance components. The brightness usually drops off with off-axis viewing as a result of some directionality in the backlight and variations in the transmittance of the LC cell with angle.

Brightness enhancement films between the backlight and display, to be discussed in Chapter 6, can further raise luminance at normal viewing, sometimes at the expense of off-angle luminance. This is done for notebook panels to raise luminance for single viewers without increasing power consumption in the backlight.

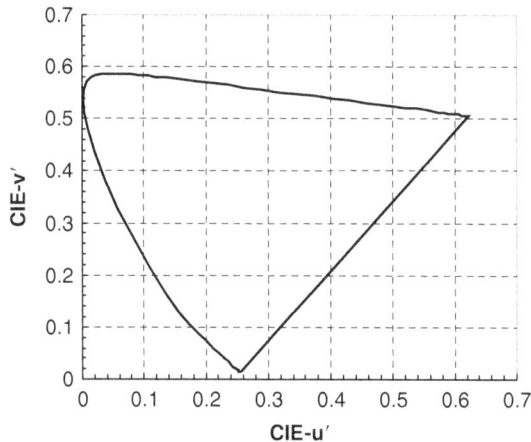

Figure 5.3: CIE 1976 u'v' Chromaticity Diagram with uniform JNDs.

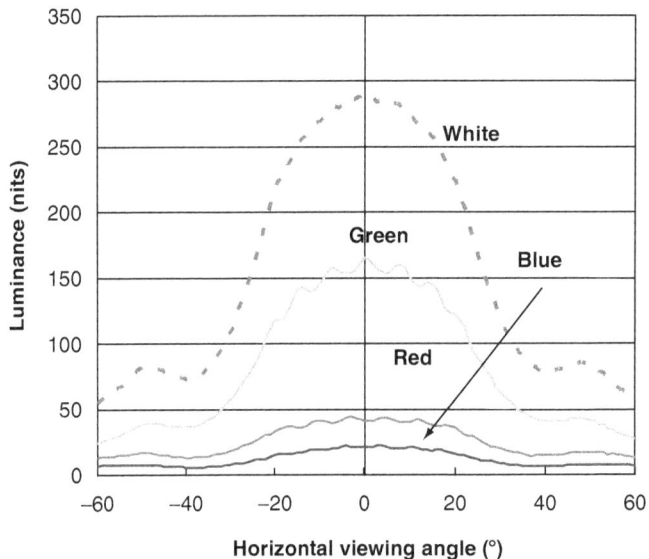

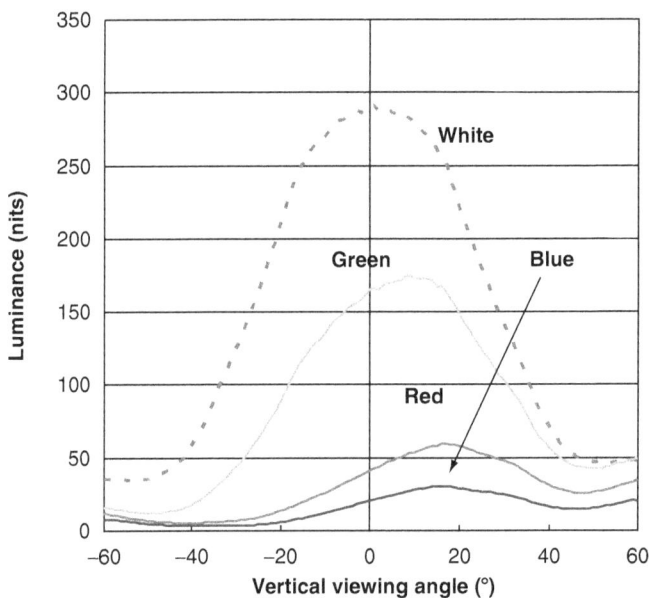

Figure 5.4: Dependence of luminance on viewing angles in a typical TN AMLCD.

Besides the backlight, the components of the assembly determine the final display luminance (Fig. 5.5). Starting with a luminance of 3000 cd/m², the final white luminance from the display is only about 150–300 cd/m² because of losses in the LCD. The first polarizer transmits about 42% of the light. The black matrix opening on each pixel (the aperture ratio or fill factor) is around 60% and the average color filter transmittance is about 25% of incident white light. Other layers and the exit polarizer further reduce luminance so that total transmittance of a backlit AMLCD is only 5–10%.

The contrast ratio of the display is defined as the ratio of maximum to minimum luminance (full white to full black state):

$$CR = \frac{L_{max}}{L_{min}}. \qquad (5.3)$$

CR can exceed 500:1. For the TN cell of Fig. 5.6, the contrast ratio drops off rather quickly with off-angle viewing. To maximize peak contrast ratio, it is important that the polarizer placement and the rubbing directions of the alignment polyimide are accurately controlled in manufacturing. If this is the case, the contrast ratio of the display at normal viewing angle can approach the extinction ratio of the polarizers.

For the TN cell it has been shown that with zero applied voltage, the transmittance $T_{NB,V=0}$ in the normally black (NB) mode and $T_{NW,V=0}$ in the normally white (NW) mode depend on wavelength according to the Gooch–Tarry theory [4]:

$$T_{NB,\,V\,=\,0} = T_{pols} \frac{\sin^2\left(0.5\pi \sqrt{1 + u^2}\right)}{1 + u^2}, \qquad (5.4)$$

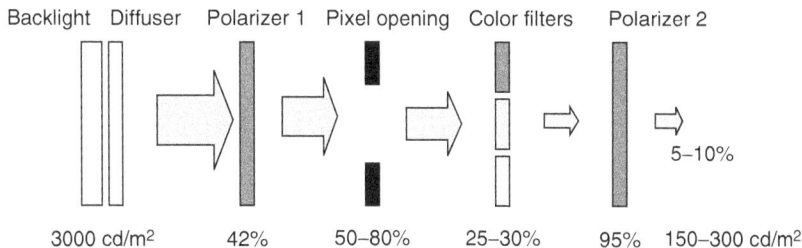

Figure 5.5: Transmission of various components in an AMLCD resulting in 5–10% overall transmission.

$$T_{NW, V = 0} = T_{pols}\left|1 - \frac{\sin^2\left(0.5\pi\sqrt{1 + u^2}\right)}{1 + u^2}\right|, \tag{5.6}$$

$$u = \frac{2\Delta n.d}{\lambda}. \tag{5.7}$$

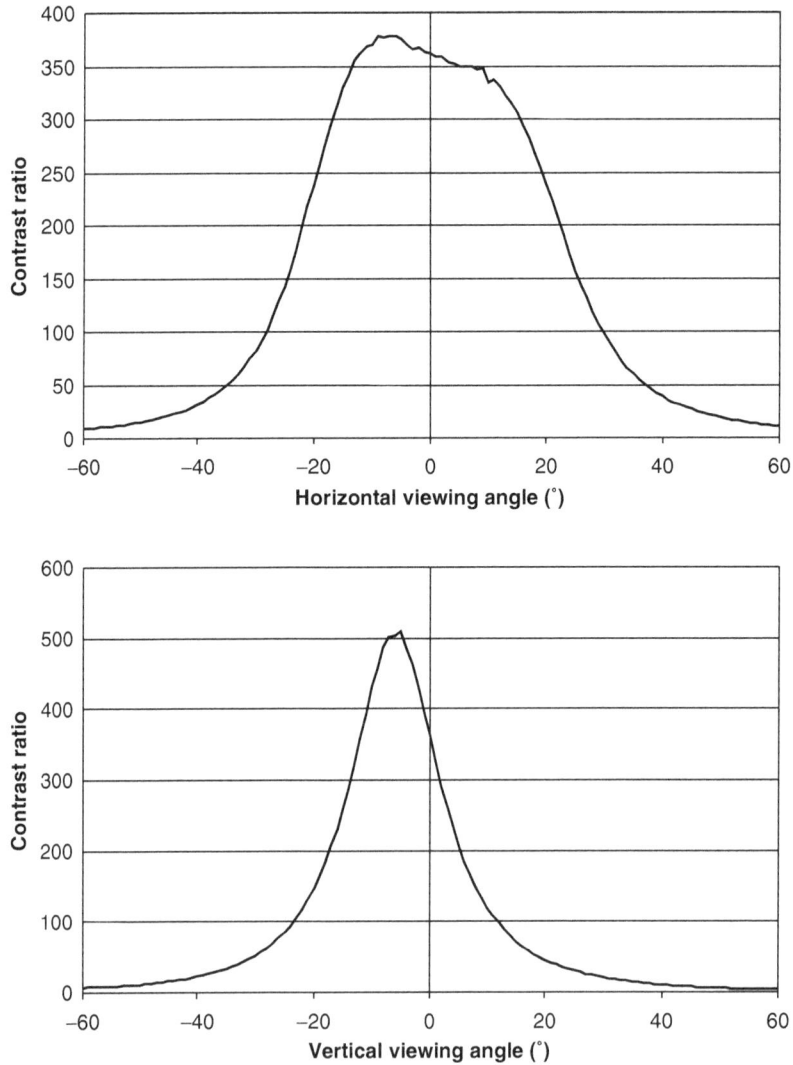

Figure 5.6: Dependence of contrast ratio on viewing angle in a typical TN AMLCD.

The parameter u is called the retardation index and T_{pols} is the transmittance of two parallel polarizers ($\approx 42\%$). Δn, d, and λ are the LC optical anisotropy, the LC cell thickness, and the wavelength of light, respectively.

The transmittance for both modes is graphically shown in Fig. 5.7. There are minima in the NB OFF transmittance for $u = \sqrt{3}$ (the first minimum), $\sqrt{15}$ (the second minimum), $\sqrt{35}$ (the third minimum), etc. In the NB mode, high contrast ratio can only be obtained for monochromatic light with a wavelength corresponding to one of the minima.

Since LCDs normally operate with white light, the OFF state luminance in the NB mode will be significant when integrating over the spectrum. This is the reason why the NB TN mode cannot easily achieve high contrast ratio and is seldom used. In the NW mode, the OFF state luminance (dark state) is obtained at high applied voltage (see also Fig. 1.11 in Chapter 1) and is low and less dependent on wavelength. The NW mode can therefore achieve high contrast ratios exceeding 500:1 for a narrow viewing cone, as shown in Fig. 5.6.

5.3 Viewing Angle Behavior

The transmission-voltage curves of TN cells vary significantly with viewing angles (Fig. 5.8). Particularly at intermediate gray levels, the curves for the vertical viewing angle diverge strongly.

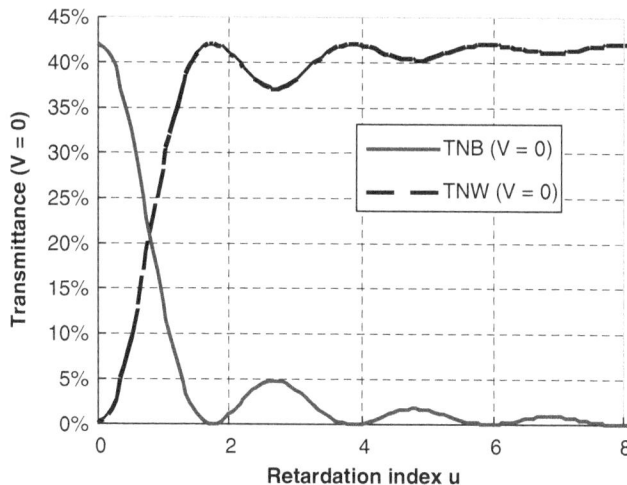

Figure 5.7: Transmittance of TN cell without applied voltage in NB and NW mode as a function of wavelength ($u = 2d.\Delta n/\lambda$).

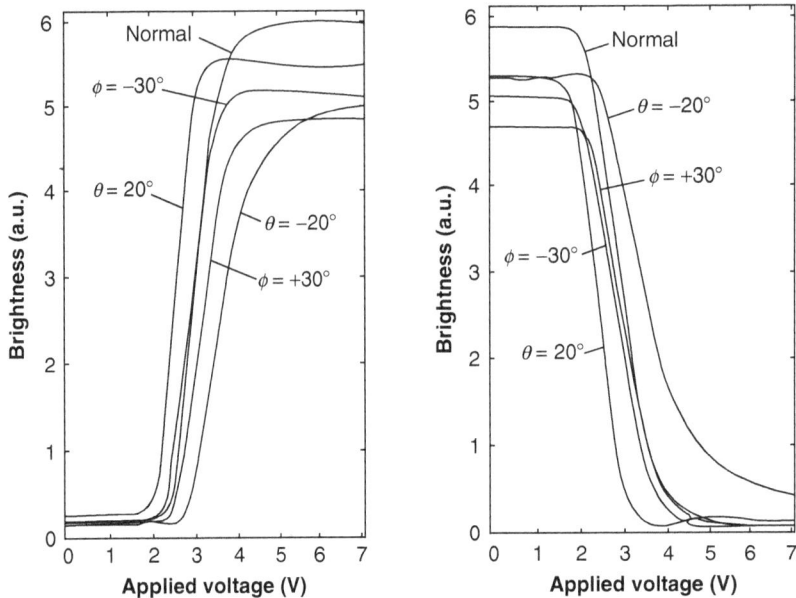

Figure 5.8: Dependence of TN cell transmittance-voltage curves on viewing angle in NB mode (left) and NW mode (right). φ is the horizontal viewing angle and θ is the vertical viewing angle.

This can be explained by the orientation of the LC molecules in the center of the cell, which is very different for positive and negative vertical viewing angles. In Fig. 5.9 this is illustrated in a top view and cross section of a standard TN cell. The surfaces are rubbed at +45 degrees and −45 degrees to obtain corresponding alignment of the LC molecules. This causes the horizontal viewing cone to be symmetric. When a voltage larger than the TN threshold voltage is applied, the LC molecules tilt parallel to the vertical plane so that the retardation is quite different for upper and lower viewing directions.

This explains why the transmittance curves vary strongly with vertical viewing angle, in particular at mid-gray levels. Around these voltage levels the LC molecules are only partially tilted in the center of the cell. The viewer will see a very different orientation of the molecules (and therefore cell transmittance) when viewing from the lower or upper viewing angle.

The transmittance-voltage curves of Fig. 5.8 also show that at some vertical viewing angles gray scale inversion in the NW TN cell occurs (i.e., the transmittance actually increases with increasing LC voltage). This leads to the negative gray scale image observed in most notebook LCDs when viewed from some lower oblique vertical angles.

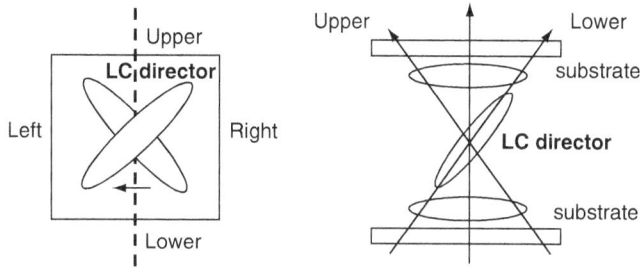

Figure 5.9: Top view (left) and cross section of a TN LC cell explaining the strong variation in viewing from upper and lower angles.

The normally black TN mode has better contrast ratio at off-angle viewing than the normally white mode, but a lower peak contrast ratio with a white backlight, as explained before.

Many techniques have been introduced during the past ten years to improve on the poor viewing angle of the TN cell; the most important methods are described in Chapter 6.

5.4 Color and Gray Scale Performance

The backlight in active matrix LCDs usually has cold cathode fluorescent lamps (CCFLs), placed either at the edge of the panel (in notebook panels and most monitors) or behind the panel (for LCD television and some monitors). The CCFL backlight consists of one or more stick lamps coated on the inside with red, green, and blue phosphors. The discharge in the lamps leads to emission of white light having a spectrum with three distinctive peaks in red, green, and blue (Fig. 5.10). The peaks correspond to the characteristic emission spectrum of the phosphors.

The individual subpixels in the display have red, green, and blue color filters. Color representation is based on the additive color principle: at normal viewing distances the observer will effectively perceive the mix of colors from the three subpixels making up each pixel. When all three color subpixels are transmitting, the pixel color will be perceived as white. The exact color coordinates of the white point (the color temperature of the display) depend on the relative transmission and color purity of the red, green, and blue subpixels.

The red, green, and blue color filters have a spectral transmittance as shown in Fig. 5.11.

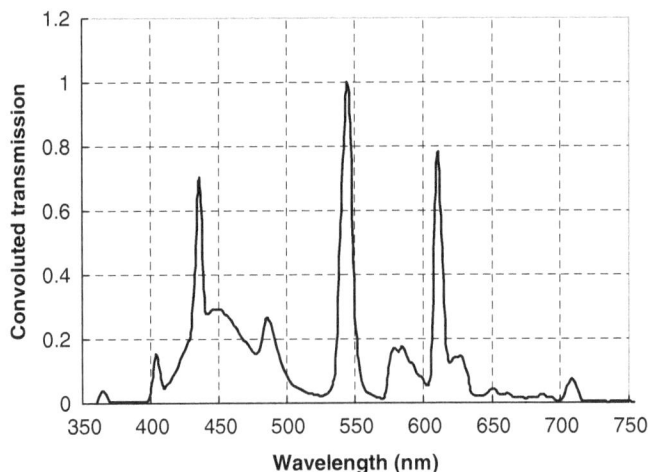

Figure 5.10: Spectrum of tri-phosphor backlight.

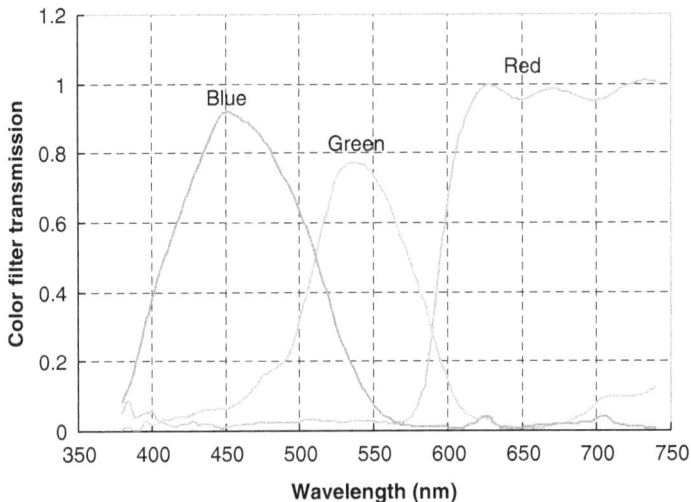

Figure 5.11: Typical spectra of red, green, and blue color filters.

Ideally, the peaks in the backlight spectrum match the color filter transmission, for efficient color rendering on the display. Assume the spectrum of the backlight is given by $S(\lambda)$ and the spectra of the color filters by $R(\lambda)$, $G(\lambda)$, and $B(\lambda)$. Assume also that the other films and layers in the display (polarizers, LC fluid, ITO layers, etc.) have a flat white transmittance spectrum. Then, for an infinite contrast ratio, the spectra of the most saturated colors in the display are given by the convolution of the color filter and backlight spectra:

$$RED = R(\lambda)S(\lambda)$$

$$GREEN = G(\lambda)S(\lambda) \quad . \tag{5.8}$$

$$BLUE = B(\lambda)S(\lambda)$$

In practice, the contrast ratio is finite for each color, which means that the colors R, G, and B of the display are slightly less saturated than in Eq. 5.8.

The resulting convoluted spectra of the primary colors are shown in Fig. 5.12.

If the convoluted spectra would have a single monochromatic wavelength, their color coordinates would lie on the horseshoe curve of the chromaticity diagram of Fig. 5.2, for a very wide color gamut. Since there are some other colors present in each of the three spectra, the color gamut is typically shrunk to, for example, the triangle shown in Fig. 5.2. The challenge of improving the color gamut is in optimizing the combination of backlight and color filter spectra.

In most AMLCDs the color filters are patterned in a vertical stripe arrangement coinciding with the subpixels on the TFT array (Fig. 5.13). This gives best performance for graphic displays in computers. The color mixing is, however, not ideal in this configuration because each color is lined up as a vertical stripe.

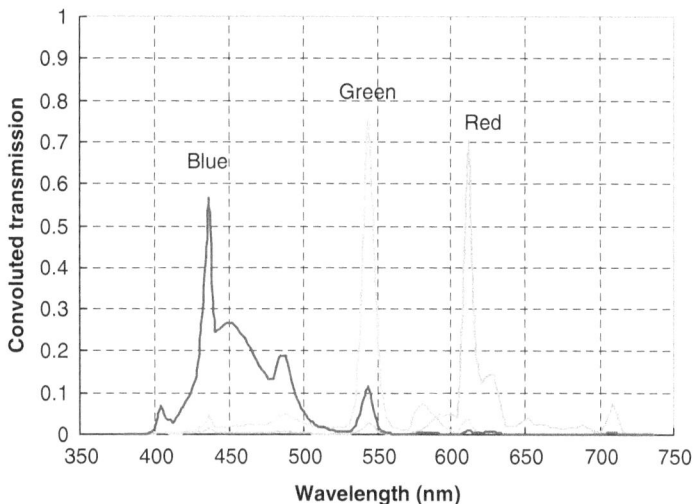

Figure 5.12: Spectra of primary colors displayed on the LCD with the backlight spectrum of Fig. 5.10 and the color filter spectra of Fig. 5.11.

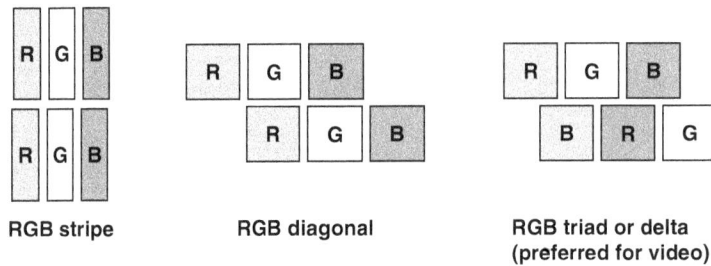

Figure 5.13: Color filter arrangements.

Alternative color filter arrangements include RGB diagonal and RGB triad or delta. RGB diagonal gives better color mixing but may require a more complicated data driver IC, in which each channel can supply different color data signals. RGB triad gives best performance for video but requires staggered routing of the data buslines on the TFT array.

The vertical stripe arrangement is attractive for another reason: It leads to half the number of rows in the display, as compared to the other configurations. This, in turn, makes it possible to keep the line select time sufficiently long so as to reduce the demands on the ON current of the TFT and the RC delays on the buslines, especially for very high resolution and large displays.

The number of colors that can be achieved with any of these color filter arrangements depends on the number of gray levels per color. When an analog data driver is used with continuously varying gray levels, the number of colors is, for all practical purposes, infinite.

For an n-bit digital data driver the number of gray levels is 2^n and the number of colors is 2^{3n}. Table 5.2 shows the results for data drivers with an increasing number of gray scale

Table 5.2: Relation between bits of gray scale, number of gray levels, and number of colors

Number of bits supplied by data driver	Number of gray levels on display	Number of colors on display
1	2	8
2	4	64
3	8	512
6	64	262,144
8	256	16,777,216
10	1024	1,073,741,824

bits. Full-color displays have an 8-bit gray scale, sufficient for full-color video. High-definition LCD television has up to 10-bit gray levels. Some special applications require even more bits of gray scale. An example is monochrome LCDs for medical imaging, which can have as many as 3061 gray levels (more than 11 bits).

The number of bits in the data driver determines how many different voltage levels can be stored on the pixel. Whether these levels can actually be distinguished depends on the gamma control of the display and the sensitivity of the human eye. In Chapter 4, gamma control circuits for implementing gray scale were introduced. In Fig. 5.14 the transmittance-voltage curve of the normally white TN cell is shown, along with eight gray level data voltages which would result in a linear dependence of the transmittance on the gray level number 1, 2,...,8, corresponding to V_1, V_2,...,V_8. As is seen from Fig. 5.14, the voltage levels to obtain a linear gray scale are not equidistant because the transmittance voltage curve of the LC cell is not linear.

The human eye's sensitivity to light is not linear, and gray-scale representation must take this into account. One popular gray scale is derived from another CIE 1976 standard, based on experiments on JNDs with human observers. It was found that the human vision response L^* to a luminance stimulus can be quantified with the following equation:

$$L^* = 116\left(\frac{Y}{Y_0}\right)^{1/3} - 16, \frac{Y}{Y_0} \geq 0.008856, \tag{5.9}$$

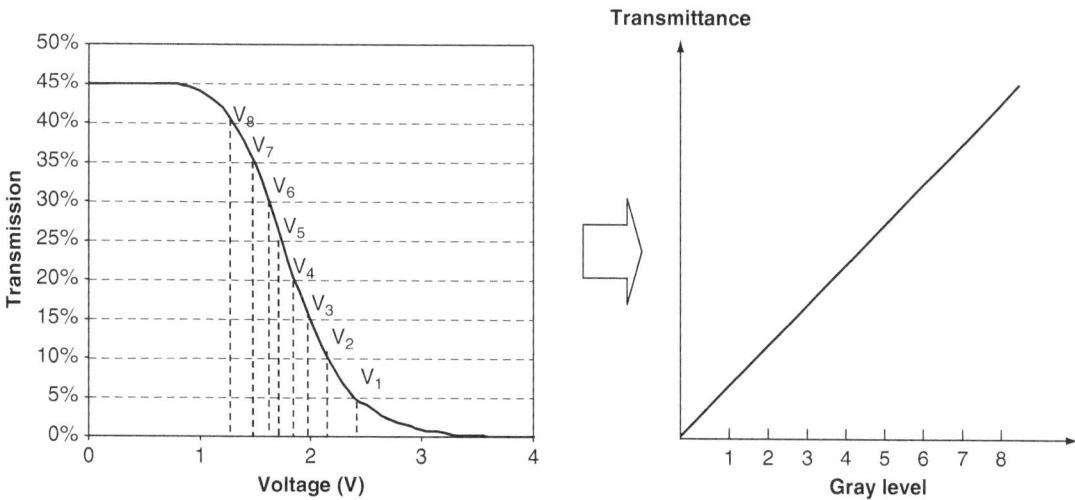

Figure 5.14: Voltage levels to obtain a linear gray scale in a normally white TN LCD.

where L^* is expressed in percent and Y/Y_0 is the relative luminance.

This relation is shown in Fig. 5.15. It implies that at low light levels smaller luminance differences can be distinguished than at high levels.

The gamma correction for the gray scale takes into account the L^* (or some other response) and the transmittance-voltage curve of the LC cell. The best gray scale is obtained when, for example, for an 8-bit gray scale the difference between each of the 256 levels is optimized and is most clearly observable. This occurs when the gamma factor γ equals 3 to compensate for the human vision response. The display luminance $L_{display}$ versus gray level number (for a 256-level gray scale) is then given by

$$L_{display} = L_{max}\left(\frac{graylevel\#}{256}\right)^{\gamma}. \qquad (5.10)$$

The relationship between display luminance and gray level number is graphically depicted in Fig. 5.16 for a linear gray scale and for the compensated gray scale.

In a display with an 8-bit gray scale (256 gray levels), the total number of colors exceeds 16 million, sufficient for full-color displays. In Fig. 5.17 the color gamut for a TFT LCD is shown in the x,y and u',v' chromaticity diagrams. The area of the color gamut is often expressed as a percentage of a standard area. For example, the color gamut of LCD desktop monitors typically covers about 70% of the NTSC standard color gamut, shown in Fig. 5.17.

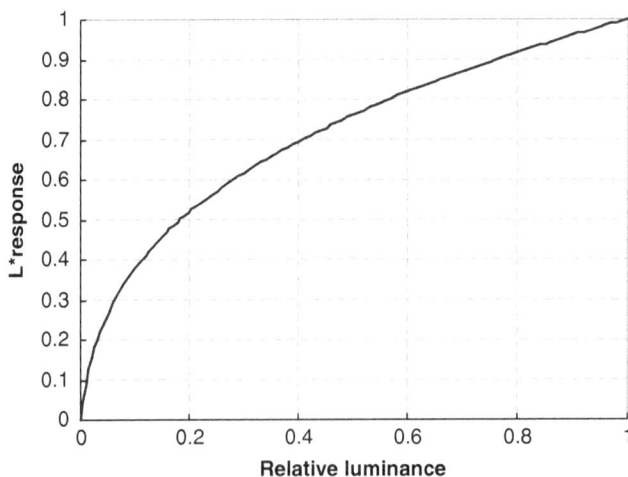

Figure 5.15: Human vision response L* versus brightness stimulus.

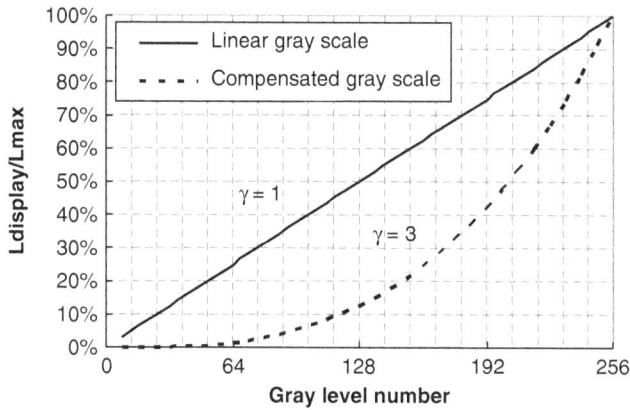

Figure 5.16: Display luminance versus gray level for a linear gray scale and a gray scale compensated for the human vision response.

A high contrast ratio ensures that the primary colors are saturated at normal viewing. With drop-off of contrast ratio at oblique angles, colors are less saturated at off-angle viewing. As a result, the color gamut will shrink for oblique angles.

Recently, combinations of color filters and backlights have been demonstrated that give 100% coverage of the NTSC gamut. Figure 5.17 also shows the white point or color temperature near the center of the gamut. The color temperature of the LCD is measured with all three colors fully transmitting.

5.5 Response Time and Flicker

Response time is important when the display needs to render moving images. In computer displays with mostly static images, response times need to be low enough to limit the annoying submarining of fast-moving cursors. Response time depends on the square of the LC cell gap and on LC parameters such as viscosity, dielectric anisotropy, and elastic constant. Typical values are 10–50 msec. The rise and fall response times τ_r and τ_f for full white-to-black and black-to-white changes, respectively, are defined as the time to transition from 10% to 90% transmittance and vice versa. For the TN cell they are given by

$$\tau_r = \frac{\gamma_1 d^2}{\varepsilon_0 \Delta\varepsilon_{LC} V^2 - \pi^2 K},$$

(5.11)

$$\tau_f = \frac{\gamma_1 d^2}{\pi^2 K},$$

(5.12)

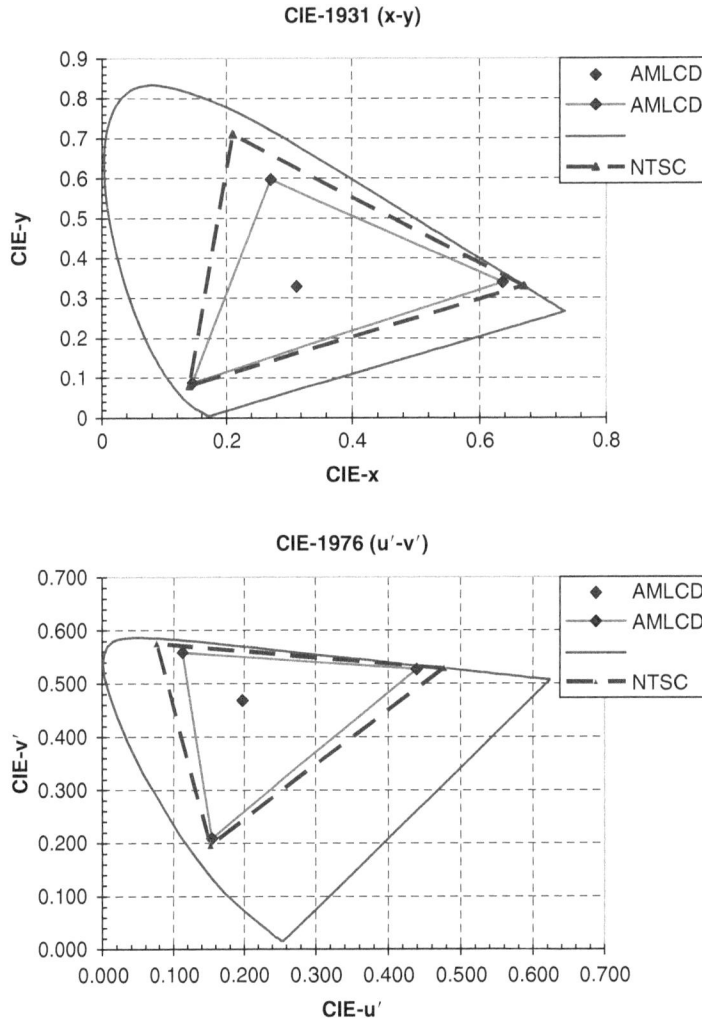

Figure 5.17: Chromaticity diagrams with NTSC standard curves, the AMLCD color gamut, and the color temperature of the AMLCD.

where γ_1 is the LC rotational viscosity, d is the LC cell gap, V is the applied voltage, K is the LC elastic constant, and $\Delta\varepsilon_{LC}$ is the LC dielectric anisotropy. In the case of the TN cell, K is a combination of the elastic constants K_{11}, K_{22}, and K_{33} for splay, twist, and bend, respectively, because all three play a role in the operation of the TN cell.

The response times are measured by periodically applying a square voltage pulse with different amplitude to the LC cell, as shown in Fig. 5.18 for a normally white TN LCD.

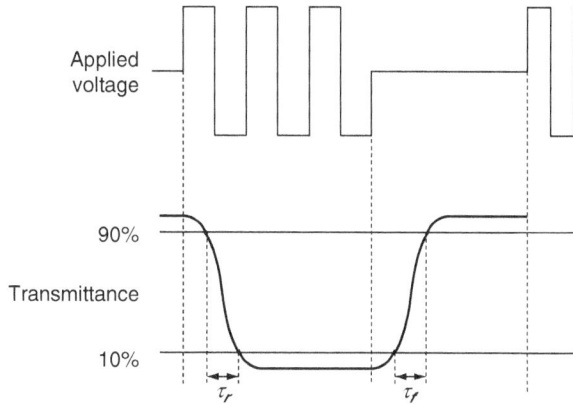

Figure 5.18: Measurement method and definition of response times in normally white TN LCD.

It should be noted that the response time between intermediate gray levels can be much longer than the response times between the black and white states, as is evident from Eq. 5.11.

A fast response time is needed to reduce smearing of video images, but it can also cause flicker unless precautions are taken. Flicker occurs when the luminance varies by more than a few percent at a frequency less than about 40 Hz, as discussed earlier in Chapter 4, Sec. 4.1. Higher frequencies of luminance variation are averaged by the human eye and are not perceived as flicker. In a display that is refreshed at a 60-Hz rate, 30-Hz flicker can occur because the total period consisting of a positive and negative voltage cycle across the LCD is 33 msec (corresponding to 30 Hz). It is difficult or impossible to completely eliminate small DC components across the LC for all gray levels and at all locations on the display. The DC component causes flicker at 30 Hz, when all pixels are driven at the same polarity in odd frames and all at the opposite polarity in even frames. This inversion method is called frame inversion and is normally avoided. In Chapter 4, line, column, and dot inversion drive methods were described to eliminate flicker.

5.6 Resolution and Size

Table 5.3 lists common display formats with their number of addressable pixels, aspect ratio, and row select time at a 60-Hz refresh rate. Each color subpixel is addressed individually and the data voltage on three subpixels of one color group determine the luminance and chromaticity of one square color pixel by additive color mixing.

The meaning of resolution versus number of pixels can be different in matrix displays, such as LCDs, and in CRTs, which are addressed by an electron beam with a finite spot

Table 5.3: Common display formats

Resolution	Number of pixels	Aspect ratio	Row select time at 60 Hz refresh rate
QCIF	144 × 176(×3) – 0.025 Mpix	3:4	100 μsec
QCIF+	220 × 176(×3) – 0.039 Mpix	4:3	70 μsec
QVGA	240 × 320(×3) – 0.08 Mpix	3:4	60 μsec
CIF	288 × 352(×3) – 0.1 Mpix	3:4	50 μsec
VGA	480 × 640(×3) – 0.3 Mpix	3:4	30 μsec
SVGA	600 × 800(×3) – 0.48 Mpix	3:4	25 μsec
XGA	768 × 1024(×3) – 0.8 Mpix	3:4	22 μsec
W-XGA	768 × 1280(×3) – 1 Mpix	9:15	22 μsec
SXGA	1024 × 1280(×3) – 1.3 Mpix	4:5	15 μsec
SXGA+	1050 × 1400(×3) – 1.5 Mpix	3:4	15 μsec
W-HDTV	1080 × 1920(×3) – 2 Mpix	9:16	14 μsec
UXGA	1200 × 1600(×3) – 1.9 Mpix	3:4	12 μsec
QXGA	1536 × 2048(×3) – 3 Mpix	3:4	10 μsec
QSXGA	2048 × 2560(×3) – 5.2 Mpix	3:4	8 μsec
QUXGA	2400 × 3200(×3) – 7.6 Mpix	3:4	6 μsec
W-QUXGA	2400 × 3840(×3) – 9.2 Mpix	9:16	6 μsec

CIF, common intermediate format; VGA, video graphics array; XGA, extended graphics array; HDTV, high definition television; Q, quarter or quad; S, super; U, ultra; W, wide.

size. In CRTs the effective resolution can be lower than the addressability, as a result of the Gaussian profile of the phosphor spot emission. The emission profiles of adjacent pixels can therefore overlap and reduce resolution, as shown in Fig. 5.19. In LCDs, on the other hand, the displayed resolution is equal to the addressability because of the sharp brightness profile of each pixel.

The resolution required for displays, including LCDs, depends on the application and the viewing distance. The acuity of the human eye for the average person is about one arc minute (1/60th of a degree). As shown in Fig. 5.20, this corresponds to a pixel pitch of 100 μm at a viewing distance of 36 cm, typical for a cell phone. Most humans would not be able to resolve a higher pixel density from that distance. For desktop computer monitors a viewing distance of 50 cm (20 in.) is more usual and would lead to a pixel density not higher than 180 ppi (pixels per inch).

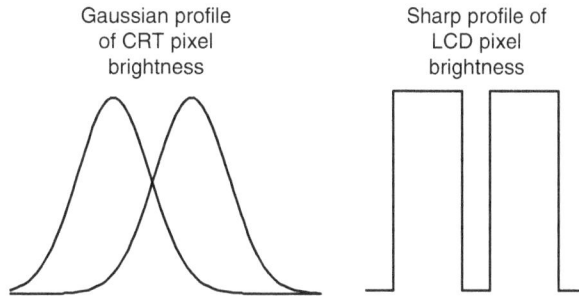

Figure 5.19: Brightness profile from CRT pixels (left) and LCD pixels (right).

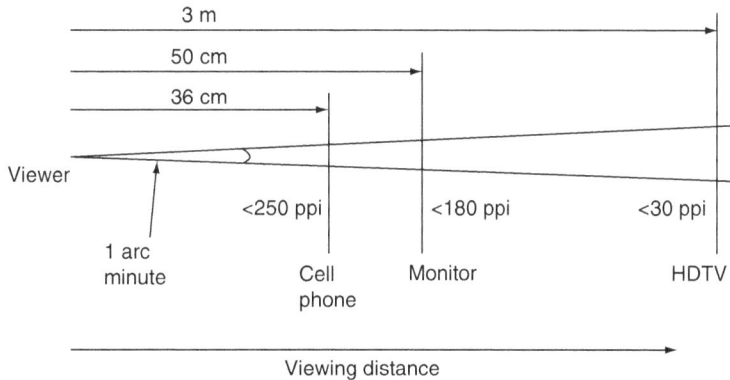

Figure 5.20: Maximum useful resolution versus viewing distance.

In the case of W-XGA LCD television viewed in a living room from a 3-meter (10-foot) distance, the maximum useful pixel density would be 30 ppi or a pixel pitch of 0.864 mm. This corresponds to a screen size for a 768×1280 pixel display of $0.65 \text{ m} \times 1.10 \text{ m}$ (50-in. diagonal). Resolutions higher than those listed in Fig. 5.20 are not very useful, since they cannot normally be distinguished by the average person.

XGA and SXGA+ resolutions with pixel pitches of 150–250 μm are typical for notebooks with 12- to 17-in. LCDs. Some high-end notebooks even have UXGA screens. Desktop monitors have 14- to 21-in. screens with XGA to UXGA resolution.

Since televisions in the home are watched from a greater distance, their resolution is lower (more in the range of VGA to W-XGA and HDTV). Television resolution is also limited by the bandwidth limitations leading to standards for NTSC (512 lines inter-laced) and HDTV (720 p or 1080 i) signal transmission.

For special applications such as medical imaging and satellite map studies, displays have a size in the 20- to 25-in. range, but are viewed up close. Such applications require the highest pixel count LCDs (in the 3 and 5 Megapixel range).

When looking at an LCD monitor larger than about 17 in. from a distance of 50 cm, the viewer sees the corners of the display at a different viewing angle than the center. Monitors are also more likely to be viewed by more than one person. This makes the simple TN cell with its strong angular dependence of the contrast ratio unacceptable. These larger displays almost exclusively use the viewing angle enhancement techniques and different LC modes described in Chapter 6.

5.7 Image Artifacts

Image artifacts on AMLCDs may be subdivided into spatial and temporal artifacts. Some of them are inherent to the technology used, while others are yield-related and can be reduced or eliminated by process optimization.

Spatial artifacts include *mura*, pixel defects, and cross-talk. *Mura*, a Japanese term for non-uniformity, applies to many types of local variations in luminance, contrast ratio, and color performance. They are usually process-related; for example, they can be caused by non-uniformities in the coating or rubbing of the LC alignment layer. Variation in the LC cell gap and contamination of the LC fluid are other causes of *mura*. For example, when the distribution of spacers controlling the cell gap is non-uniform as a result of clustering, the LC thickness will vary. This can lead to contrast ratio variations and poor dark uniformity, especially for LC modes with a strong dependence of the black state on cell gap (e.g., IPS LCDs, Chapter 6). Process and materials optimization are used to contain *mura* within the specifications of the LCD.

Pixel defects are a yield issue, and were discussed in Chapter 4. Again, design and process optimization are the key to minimizing these defects. In some cases, laser repair of pixel defects is possible. Since bright pixel defects are more objectionable than dark pixel defects, conversion of bright into dark spots by laser zapping is common for large displays.

Cross-talk is a variation in luminance of a pixel or group of pixels depending on the video data supplied to the other pixels on the same column(s) or row(s). The first is vertical cross-talk; the second is horizontal cross-talk. They usually manifest themselves by a viewing area section with a darker or lighter shade of gray next to a dark or light area. An example is shown in Fig. 5.21, where a black square on a gray background causes the

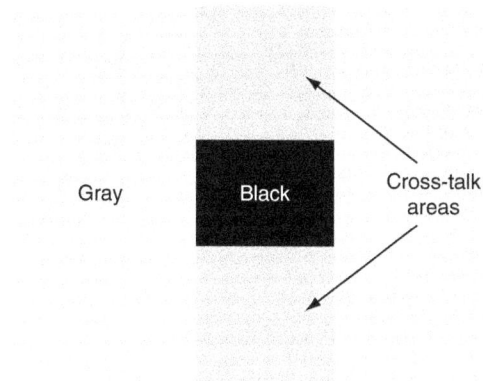

Figure 5.21: Example of vertical cross-talk.

area above and below the square to have a darker shade of gray than intended. This type of vertical cross-talk can be quite severe in passive matrix LCDs and is one of the reasons why active matrix LCDs have replaced STN LCDs for all larger and higher-resolution displays.

Cross-talk is quantitatively defined in most AMLCD specifications. It can be caused by non-optimized drive schemes, by TFT leakage in the OFF state, by RC propagation delays on the buslines, or by capacitive coupling between the pixels and buslines. Cross-talk is minimized by designing pixel layouts with sufficiently low capacitive coupling between the pixel and the row or column buslines (including from the TFT itself), by the use of low-resistance buslines and by adjusting the shape and timing of the gate and data drive wave form. As explained in Chapter 4, dot inversion driving methods help eliminate cross-talk by canceling out the capacitive coupling from negative and positive data voltages. All of these parameters are usually optimized during the design of the panel with the aid of circuit modeling.

Temporal artifacts include image retention, flicker, and motion blur. Image retention (also called image sticking, latent image, or burn-in) may be defined as a local variation in luminance depending on the image displayed in a preceding period of time. It can have different manifestations and causes, and is usually quantitatively defined in the LCD specification. Short-term image retention, which fades away after several seconds or minutes, is usually caused by residual DC components on the LC voltage that charge up the LC alignment layer. This charge gradually changes when the display data to the pixels is changed. The remedy is to make sure that the DC component does not exceed 50–100 mV at all gray levels. Selecting alignment layer materials that dissipate the charge build-up is another approach to minimize the problem.

Another source of this type of fading image retention is disclinations in the LC layer. One possible cause is the topography on TFT array, as illustrated in Fig. 5.22.

In the flat pixel area, the pre-tilt of the LC layer (1–3 degrees) at the surface with the alignment layer ensures the proper tilt of the center layer of the LC when a voltage is applied across the LC pixel. At the pixel edges the topography of the TFT structures and buslines can cause a reverse tilt domain. At the boundary of this reverse tilt domain and the domain with intended tilt on the pixel, there is a disclination where light leakage occurs. The disclination is usually located in a corner of the pixel. When the data voltage on the pixel is changed, this disclination can slowly change or disappear, causing fading image retention. The solutions are to planarize the TFT array, use a higher pre-tilt angle, or cover up the disclinations with the black matrix on the color plate (at the expense of pixel aperture). Planarization minimizes the topography on the pixel so that the LC molecules at the surface tilt in the same direction at all locations. Chapter 6 will describe high-aperture designs with planarization layers, which reduce the possibility of reverse tilt. Visible reverse tilt domains in the pixel will also affect contrast ratio and, when they vary across the viewing area, the black state uniformity.

In TN LCDs a chiral dopant is added to the LC fluid to create a preferred twist direction (see Chapter 1, Fig. 1.10). If this is missing or deliberately taken out, another source of light leakage is the boundary between domains with different twist (+90 or −90 degrees). In properly designed and manufactured LCDs this is not an issue.

Permanent burn-in or long-term image retention is quite rare in AMLCDs and can be caused by large changes in TFT ON or OFF current or by permanent changes in the alignment layer.

Flicker can be caused by luminance changes in the backlight or by DC components on the pixel voltage. Even a small DC component of 10–50 mV can cause flicker, especially at mid-gray levels. It is difficult to prevent such small DC components across the entire

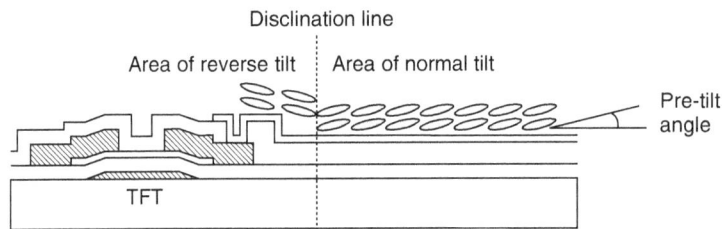

Figure 5.22: Pixel cross section showing a reverse tilt domain causing image retention and non-uniformities.

viewing area and for all gray levels. As explained in Chapter 4, inversion drive methods are used to cancel out flicker by spatial averaging.

Motion blur is caused by the slow response time of LC fluids to voltage changes and by the hold-type character of LCDs. This problem and methods to solve and minimize it will be discussed in more detail in Chapter 6, Sec. 6.6.

References

1. F. Graum and C.J. Bartleson, Eds., *Optical Radiation Measurements: Volume 2, Color Measurement*. New York: Academic Press (1980).
2. L.E. Tannas, Ed., *Flat Panel Displays and CRTs*. New York: Van Nostrand Reinhold (1983).
3. G. Wyszecki and W.S. Stiles, *Color Science Concepts and Methods*. New York: John Wiley & Sons (1982).
4. C.H. Gooch and H.A. Tarry,. "Optical Characteristics of Twisted Nematic Liquid-Crystal Films," *Elect. Lett.*, 10, 1575–1578 (1974).

6

Improvement of Image Quality in AMLCDs

Over the past 15 years, the performance of AMLCDs has dramatically improved. Depending on the requirements for a particular application, the characteristics described in the previous chapter have been optimized. They include power consumption for portable devices, viewing angle for desktop monitors and televisions, contrast ratio, brightness, color gamut, and video response times for LCD televisions. In Table 6.1 the importance of different performance characteristics is listed for various applications. For mobile applications such as notebooks, PDAs, cell phones, camcorders, and digital cameras, low power consumption is essential to prolong battery life.

For non-portable, large AMLCDs the viewing angle and response time tend to be more of a concern. Many efforts have gone into the development and commercialization of AMLCDs with application-specific optimization of the viewing characteristics. These improvements are described in this chapter.

6.1 Brightness Improvements

The brightness of transmissive LCDs depends to a large extent on the backlight intensity. The most straightforward way to increase display luminance is therefore to turn up the backlight or to design brighter backlights. The display luminance is then proportional to the backlight intensity and can reach more than 1000 cd/m^2. However, there are drawbacks to raising backlight luminance. First, it increases power consumption, which is undesirable (particularly for portable applications). Second, it may require a different construction of the backlight (for example, multiple lamps rather than single lamp edge lighting) and can also have a negative impact on backlight lifetime. Third, high backlight intensity can cause heating of the display beyond 40–50°C, where display image quality may start deteriorating. Alternatives to raising the backlight intensity are addressed in the following subsections.

Table 6.1: Importance of AMLCD characteristics for various applications

Parameter	Small mobile	Laptop	Desktop	Television
High brightness	x	x	Δ	o
Wide viewing angle	x	x	o	o
Wide color gamut	x	Δ	Δ	o
Fast video response time	Δ	Δ	Δ	o
Sunlight readability	o	Δ	x	Δ
Low power consumption	o	o	x	x
Low weight	o	o	x	x
Thin profile	o	o	Δ	Δ

o, very important; Δ, important; x, less important.

6.1.1 Increased Color Filter Transmission

Besides the backlight, the components of the assembly determine the final display luminance, as shown in the previous chapter in Fig. 5.5. The first polarizer transmits about 42% of the light. The black matrix opening on each pixel (the aperture ratio) is typically 40–70% and the color filters transmit about 25% of white light (averaged over the red, green, and blue filter). Other layers and the exit polarizer further reduce luminance so that total transmittance is only 5–10%. It is difficult to increase the transmittance of the polarizers or color filters without affecting contrast ratio or color saturation, respectively. There is a trade-off between color gamut and brightness. By reducing the density of color pigments in the color filter materials, their transmittance at fixed thickness can be increased. This has an effect similar to reducing the thickness of the color filters: it increases the spectral width of the color filter and reduces color gamut and color saturation.

The trade-off is illustrated in Fig. 6.1, where the filter curves and chromaticity coordinates labeled 2 have twice the filter thickness or twice the pigment density as compared to those labeled 1. Thinner color filters are common in mobile phone displays where power consumption is more important than a wide color gamut. Since the color filter spectra become wider for thinner filters, display luminance improves at the expense of color purity. In Fig. 6.1B, the resulting color gamuts are schematically compared.

6.1.2 High-Aperture Ratio Designs

Efforts to improve AMLCD transmittance have focused on pixel aperture ratio. In conventional designs (Fig. 6.2A), the aperture is limited to less than about 55% because the

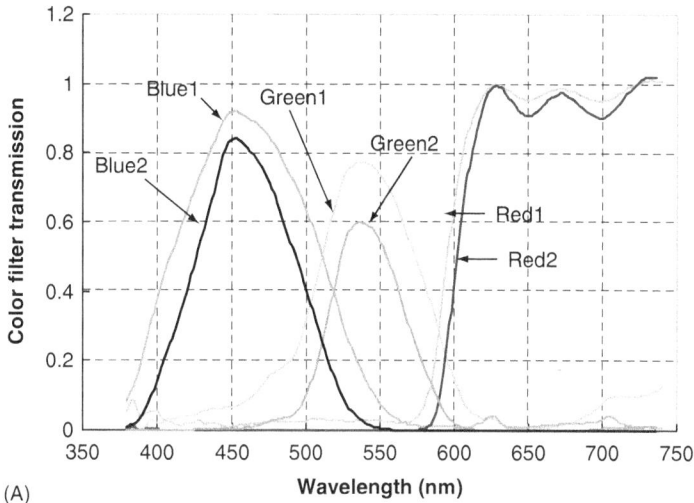

(A)

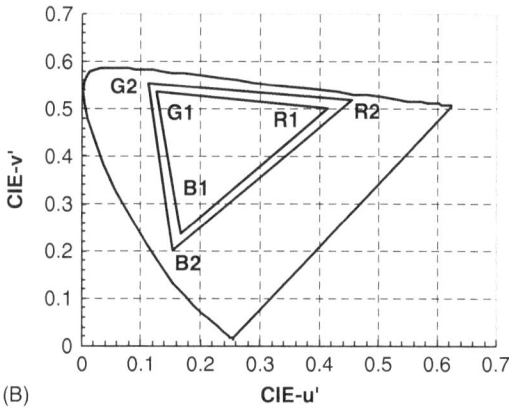

(B)

Figure 6.1: Effect of color filter thickness or pigment density on color filter spectra and color gamut.

black matrix on the color plate needs to overlap the ITO pixel electrodes on the TFT array by at least 4 or 5 μm. This is to allow for tolerance in the color-plate-to-active-plate alignment during manufacturing and to prevent light leakage at pixel edges.

In the black-matrix-on-array design of Fig. 6.2B, stripes of gate metal are added at the edge of the ITO pixel and determine the pixel opening. They block light entering the edge of the pixel under an angle. Since the registration of the layers on the TFT array to each other is very accurate (better than 1 μm), this makes a larger aperture of around 65% possible.

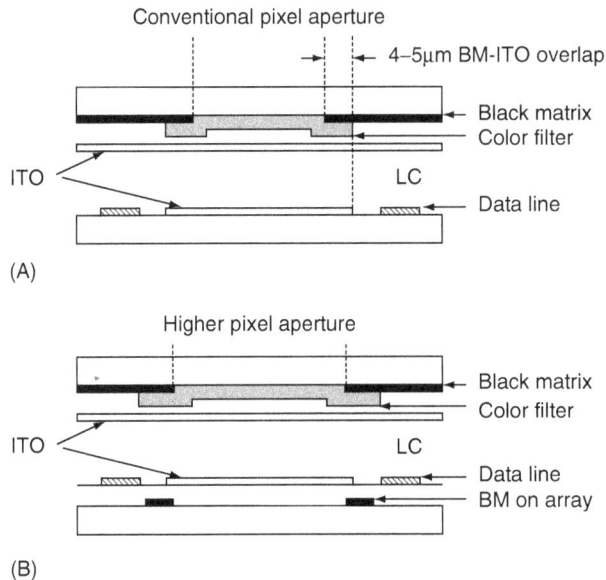

Figure 6.2: Pixel opening in conventional and in black-matrix-on-array designs.

In super-high-aperture designs (Fig. 6.3), the ITO pixel electrode overlaps the data and select buslines so that the buslines effectively function as the black matrix [1]. This leads to an even higher aperture of up to 80%. A thick polymer interlevel dielectric is used in super-high-aperture designs to passivate the TFT array. The ITO pixel electrode is patterned on top of the transparent polymer and makes contact to the TFT and the storage capacitor through vias in the polymer. The polymer is quite thick (2–3 μm) and has a relatively low dielectric constant of 2.7–3.5 to minimize the capacitance between the overlapping pixel electrode and buslines, required to keep vertical cross-talk in check.

The polymer has the added benefit of planarizing the TFT array surface and minimizing the LC disclinations at pixel edges, discussed in Chapter 5, Sec. 5.7. In principle, the black matrix becomes redundant and can be left out. However, this would expose the light-sensitive channel of the a-Si TFT to ambient light and would also cause the ambient light to reflect off the metal buslines, increasing display reflectance. An opaque, low-reflectance black matrix layer on the color plate is therefore still needed above the TFT channel as a light shield to prevent photo-leakage currents in the TFT and to minimize display reflectance. In a super-high-aperture design, however, the pixel opening in the black matrix can be much larger, as shown in Fig. 6.4. Obviously, the aperture ratio also depends strongly on the design rules and the pixel size. For larger pixels and more aggressive design rules, it tends to increase.

Super-high aperture

Black matrix
Color filter

ITO

LC

Polymer

Data line

(A)

ITO

LC

Planarized
surface

Photo-imageable polymer

TFT

Storage capacitor

(B)

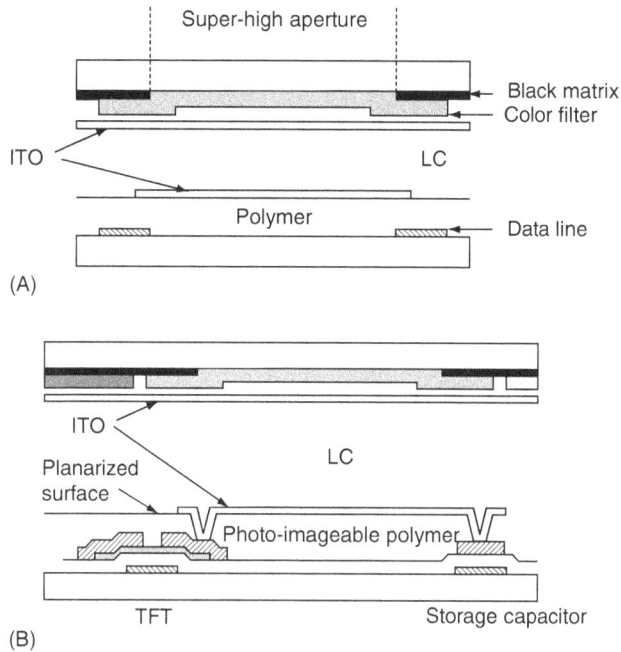

**Figure 6.3: Cross section of super-high-aperture pixel
through the data buslines (A) and through TFT and
storage capacitor (B).**

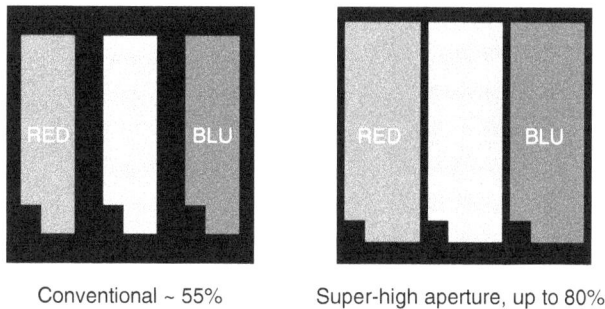

RED   BLU

RED   BLU

Conventional ~ 55%

Super-high aperture, up to 80%

**Figure 6.4: Comparison of aperture ratios in
conventional and super-high-aperture ratio pixel
designs as observed on a close-up view of an AMLC.**

High aperture is important for portable applications, allowing the backlight intensity to
be lowered while maintaining display luminance. The reduced power consumption in the
backlight extends battery life. Alternatively, a higher brightness can be achieved without
an increase in power consumption when the backlight intensity remains constant.

The material for the polymer in super-high-aperture designs may be the same as that used for the color filters except that the color pigments are not present. For example, it can be a photodefinable clear acrylic. Like the color filters, it is photo-imageable (i.e., it does not need photoresist coating, developing, and stripping, but is in itself a resist material).

Expanding on this concept, it is logical to go one step farther by simply replacing the polymer dielectric with color filter materials on the TFT substrate. This is called the color-filter-on-array (COA) method; see Fig. 6.5.

By patterning the color filters on the TFT array, the highest possible aperture is achieved (Fig. 6.5). The transparent polymer is replaced by separately patterned red, green, and blue color filters in each subpixel [2]. In the COA method the top plate has only a blanket transparent ITO layer. If the color filters cover the TFT channel they can also block most of the ambient light, eliminating the need for a black matrix. Since the common ITO layer is not patterned, there is no longer any critical alignment between the two glass plates and aperture is maximized. For very large substrates in Generation 6 and 7 facilities, it is very difficult to maintain accurate overlay accuracy between the top plate and the TFT array plate for all displays on the substrate. The COA design eliminates this problem and has therefore been considered in these advanced facilities. The drawback of the COA method is that the manufacturing complexity of the active substrate is significantly increased with at least two extra deposition and patterning steps, which can have an impact on yield.

6.1.3 Alternative Color Filter Arrangements

There are other methods to increase brightness without turning up the backlight. One that has received a lot of attention is alternative color filter arrangements. For example,

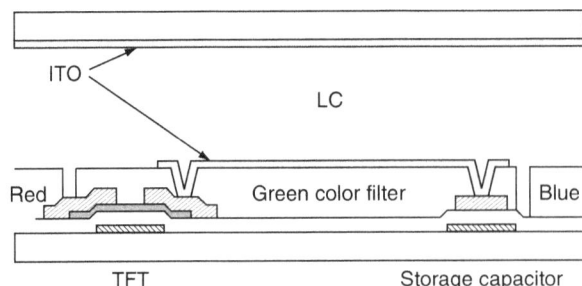

Figure 6.5: Cross section of a pixel in a COA (color-filter-on-array) TFT.

144

the use of four subpixels with red, green, blue, and white color filters was proposed and implemented in avionic TFT LCDs more than ten years ago and has now also been proposed for commercial products [3]. The white subpixel serves to increase the maximum luminance by almost 50%. It can be separately patterned as a transparent polymer without color pigment. Since color filter factories are normally set up to pattern only three color filters, the white subpixel can be left open without any material. This would make the LC cell gap larger in the white subpixel by about 2 μm, which can lead to contrast ratio problems. A blanket transparent polymer overcoat, not requiring pixelized patterning, is therefore used to planarize the color filter plate and make the cell gap more uniform.

The four subpixels can be arranged in a vertical stripe or in a quad structure, as shown in Fig. 6.6. To maximize aperture ratio, an RGBW quad arrangement is preferred, which also has the advantage of reducing the number of data driver ICs [3]. A drawback of an RGBW quad is the doubling of the number of rows and reduction of the line select time by 50%, making the quad arrangement more susceptible to propagation delays on the gate lines. This is only an issue for large and high-resolution displays. Special image processing algorithms are needed to convert RGB video signals to RGBW signals while maintaining the gray scale gamma. The overall color gamut can be slightly reduced with this approach. When the RGBW subpixel configuration is combined with a high-aperture design, the transmittance of the AMLCD can approach 15%.

6.1.4 Brightness Enhancement Films

Display screens in notebook computers and many other portable devices are generally viewed by a single person at an angle close to normal to the screen. The image quality and brightness at oblique angles is therefore less important in these applications. The

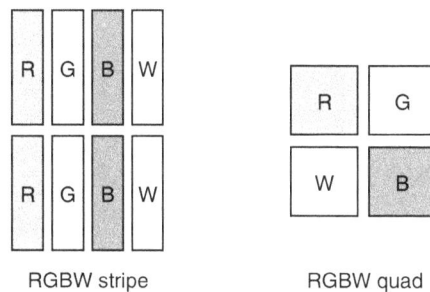

RGBW stripe   RGBW quad

Figure 6.6: RGBW color filter arrangements to improve brightness.

145

brightness at normal angles can be enhanced without increasing power consumption in the backlight by adding brightness enhancement films (BEF™) between the backlight and the display [4]. In the simplest configuration, the enhancement comes at the expense of a luminance loss at some oblique angles. The BEF™ film, produced by 3M, is basically a transparent prismatic film made from plastic, as shown in Fig. 6.7. It redirects the diffuse light from the backlight by refraction. The prisms have, for example, an angle of 90 degrees and a pitch of 20–50 µm. Some of the light is refracted toward the viewer and some is reflected back to be recycled in the backlight and reused. By using two BEF™ films (in notebooks, cell phones, and PDAs), even higher luminance can be achieved. The result is an increased brightness at normal angles of up to a factor of two.

More expensive brightness-enhancement films include polarization recovery. They transmit light of one polarization direction and reflect the other polarization component back into the direction of the backlight. The reflected component loses its polarization in the diffuse reflector of the backlight, so that it is recovered and can contribute to display luminance after a single or multiple reflections. This DBEF™ film [4] can increase luminance by up to a factor of four. Its manufacturing cost is quite high, however, so that it is used mostly for high-end applications.

In Fig. 6.8 an example is shown of the brightness enhancement obtained with various stacks of 3M's Vikuity™ films. The best performance in the figure is for a configuration which includes an enhanced specular reflector (ESR) behind the backlight.

6.2 Readability Under High Ambient Lighting Conditions

A unique feature of LCDs is their potential for very low reflectance as compared to emissive displays such as CRTs, plasma displays, and electroluminescent displays. This makes the LCD the display of choice when sunlight readability is required in high-end or out-

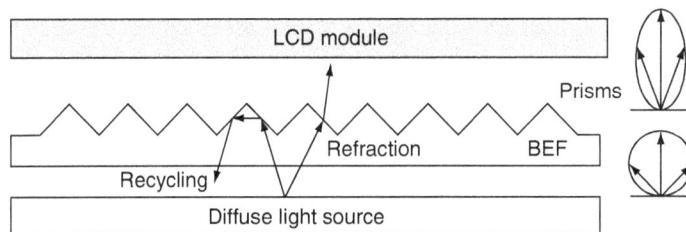

Figure 6.7: Light path in 3M's prismatic brightness enhancement film (BEF™).

door applications. However, this feature is not automatic; it needs some special steps in the manufacturing process. Without these special steps, specular reflectance is not particularly low (Fig. 6.9A). A Cr black matrix reflects about 6% and the polarizer about 7%, for a total reflection of more than 13%.

The effective contrast ratio CR_{eff} under ambient lighting is significantly reduced when there are reflections from the display:

$$CR_{eff}(A_0) = \frac{I_0 + RA_0}{(I_0/CR_0) + RA_0},$$ (6.1)

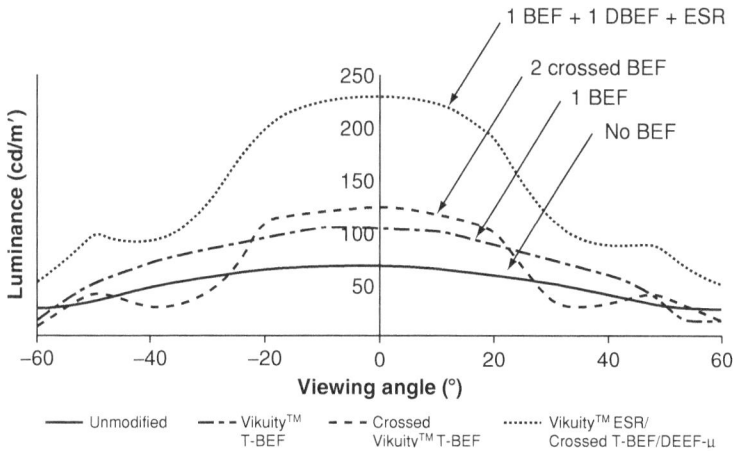

Figure 6.8: Example of brightness enhancement for different film stacks of 3M Vikuity™ films.

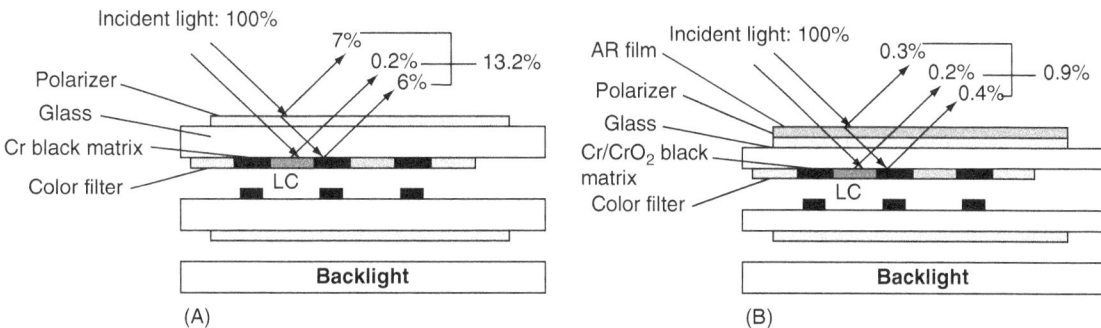

Figure 6.9: Conventional AMLCD (A) and AMLCD with minimized reflectance by adding AR film on a polarizer and by use of low reflectance Cr/CrO2 black matrix (B).

where A_0 is the ambient light intensity, I_o is the transmitted light intensity for a fully white driven area on the display, R is the reflectance of the display (which is independent of the display image content in a first approximation), CR_0 is the contrast ratio in the dark, and I_0/CR_0 is therefore the luminance transmitted for a fully OFF dark pixel in the absence of ambient light.

R can be the specular reflectance causing glare, in which case CR_{eff} is the specular contrast ratio under an angle complementary with the incident angle. R can also be the diffuse reflectance, in which case CR_{eff} is the diffuse contrast ratio.

The reflected ambient light RA_0 is approximately the same in the ON state and OFF state. It has no information, it competes with the light transmitted through the display containing information, and therefore it reduces contrast ratio and causes glare. In Fig. 6.10 the effective contrast ratio CR_{eff} is plotted as a function of the ambient light/display luminance ratio A_0/I_0 for different values of the reflectance R. As shown in the figure, the contrast ratio decreases asymptotically to 1 for high ratios of ambient lighting to display luminance.

In AMLCDs without special precautions, as well as in emissive displays, the display reflectance is typically more than 5%, so that contrast ratio drops off rapidly in bright outdoors conditions. For example, for a display with a dark room contrast ratio $CR_{eff} = 300$, the effective contrast ratio $CR_{eff} = 20$, when $I_0 = 1000$ cd/m² , $R = 5\%$ and $RA_0 = 50$

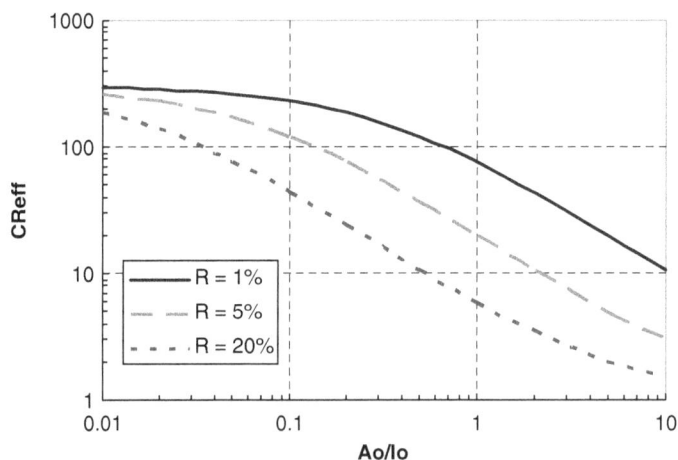

Figure 6.10: Effective contrast ratio versus relative ambient light level for different values of display surface reflections.

cd/m^2 (see Eq. 6.1 and Fig. 6.10). A high display reflectance also explains the annoying glare on emissive displays (such as plasma displays) when they are viewed in a bright environment.

In backlit AMLCDs a high contrast ratio can be maintained even at high ambient lighting by increasing I_0 and minimizing R. The unique opportunity exists to reduce the reflectance R to less than 1%. This is possible because the refractive index of the color filters is almost the same as that of glass (n = 1.5), so that reflections at the glass-color filter interface are negligible.

By using a CrO_2/Cr black matrix with 0.4% reflectance or a black polymer black matrix with even lower reflectance and an AR coating on the polarizer, total reflectance in TFT LCDs can be dramatically reduced to less than 1% (Fig. 6.9B). The glare can be further reduced by an anti-glare treatment on the front polarizer, as shown in Fig. 6.11. The surface is roughened so that incident light is scattered. This is a standard process on notebook LCDs and LCD monitors. In LCD televisions an additional AR coating on the glare-treated surface optimizes ambient light performance. The reflectance of less than 1% in LCD televisions compares favorably to around 12% in CRT television and 8% in plasma displays.

Taking these precautions makes the AMLCD legible even under bright sunlight conditions, provided that the backlight intensity is sufficient to overcome the reflected light. For outdoor applications it is recommended to avoid anti-glare coatings, and use only an AR coating on a smooth surface. The diffused ambient light outdoors can be as high as 10,000 fC, which will add a large RA_0 factor and significantly reduce CR_{eff}.

6.3   Color Gamut Improvements

The color gamut shown in Fig. 5.2 in Chapter 5 limits the number of reproducible colors to the triangle derived from the three primary colors. As explained in Sec. 6.1.1. in this

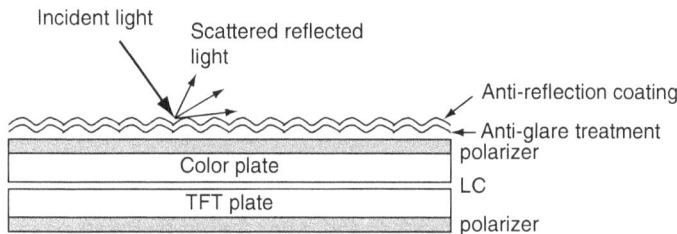

Figure 6.11: Reduction of glare by anti-glare treatment.

chapter, more saturated colors can be obtained with denser color filters, at the expense of luminance. A more attractive approach is to use a backlight with narrower peaks matching color filters with equally narrow width. For example, a well-designed LCD with an RGB LED backlight can have a better color gamut (close to 100% NTSC).

Another method to increase color gamut is the use of more than three primaries. LCDs with four and six primary colors have been demonstrated, by employing additional subpixels with different color filters. This approach requires extra processing and conversion of RGB data signals to multiple color signals in image processing. An example of the resulting improvement possible in the color gamut is schematically shown in Fig. 6.12 for a display with six primary colors. The addition of more color subpixels, which each transmit a smaller fraction of the backlight intensity, is likely to reduce luminance. Whether this color improvement is practical in terms of cost therefore depends on the application.

6.4 Wide Viewing Angle Technologies

An often-quoted drawback of conventional TN LCDs has been their poor viewing angle behavior. When observed straight on, contrast ratio and color gamut can be excellent. When the viewer turns his or her head and looks at a notebook display under an oblique angle, the contrast ratio drops fast, colors become less saturated and, under some angles, image inversion can even occur. This viewing angle behavior is usually acceptable for devices smaller than 15 in. with a single user, such as notebooks, PDAs, and cell phones.

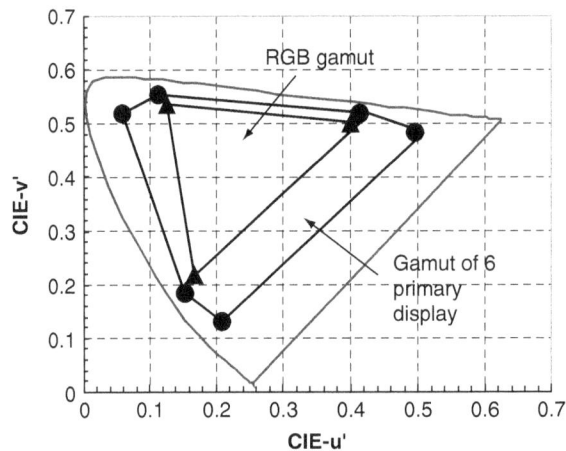

Figure 6.12: Increase in the color gamut by using more than three primary colors.

The origin of the problem was explained in Chapter 5, Sec. 5.3. and illustrated in Fig. 5.9. It can be traced back to the variation of the viewing angle relative to the LC director.

For larger, conventional TN LCDs, such as 17-in. and larger desktop monitors or LCD televisions, a single viewer at a distance of 10 or 20 in. will look at different corners of the screen under different viewing angles and the variation of contrast with viewing angle becomes objectionable. When there are multiple viewers looking at the display from different angles, similar image quality deterioration occurs.

A number of dramatic improvements in viewing angle have been developed over the past 10 years. The most important ones are:

- The addition of retardation compensation films,

- The in-plane-switching (IPS) LC mode, and

- The multidomain-vertical-alignment (MVA) LC mode.

6.4.1 Compensation Films

Retardation or compensation films are added outside the display glass assembly. They are attractive because they do not require a change in LCD manufacturing up to the final lamination of the polarizers, and they maintain the pixel aperture ratio of the original panel. Retardation films compensate for the variation over angles in the phase difference between the two orthogonally polarized components of the light wave in the LC layer.

They improve horizontal and, to a lesser degree, vertical viewing angle in TN mode LCDs. They are also used in some LCDs operating in different modes. Basically, the contrast ratio (the ratio of white to black luminance) is improved over viewing angle. Since white luminance is mostly limited by the backlight, the key to improvements is to reduce luminance in the dark state over angles.

An example is shown in Fig. 6.13 for the normally white TN cell which uses an LC fluid with positive optical anisotropy. Two so-called discotic retardation films are added outside each of the glass substrates. The films have a negative optical anisotropy and a gradually tilted optical axis. In the dark state of the normally white TN cell (with voltage applied), the LC director is similarly gradually tilted from the surface to the center of the cell. As shown in the figure, a pair-wise cancellation of the retardation in the LC fluid with positive Δn and the retardation in the film with negative Δn occurs. The total retardation of a light ray traversing the TN cell and compensation films becomes independent or less dependent on viewing angle, as shown in Fig. 6.13. The result is a significant reduction of dark luminance over wide viewing angles and a better contrast ratio over viewing angle. The discotic negative birefringence film was developed and commercialized by Fuji Photo Film® [5] and has been widely used on smaller LCD monitors.

This basic concept has been implemented with a variety of compensation films on both sides or on one side of the cell.

6.4.2 In-Plane-Switching Mode

Arguably, the best viewing angle is achieved in a different LC mode, the in-plane-switching (IPS) mode (Fig. 6.14). In the IPS mode the data voltage is applied between electrodes on

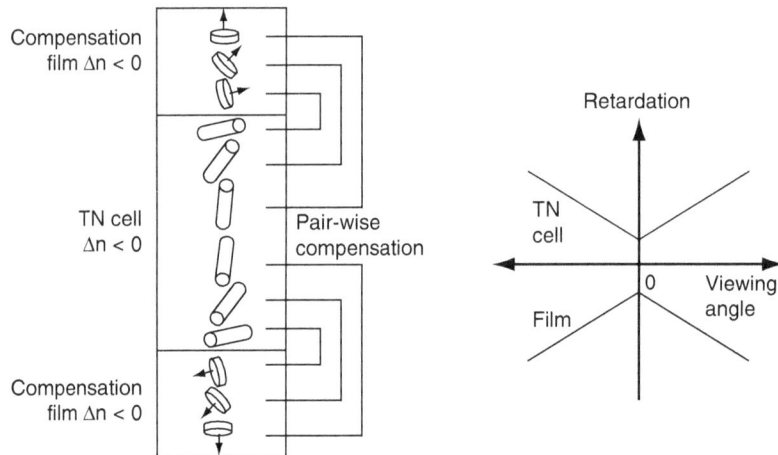

Figure 6.13: Example of a TN cell combined with compensation films and resulting retardation versus viewing angle.

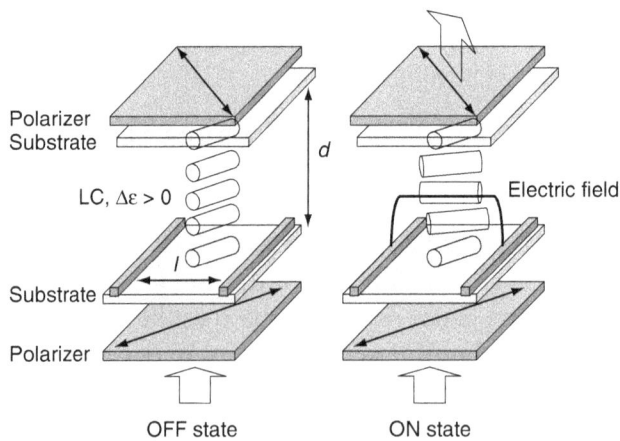

Figure 6.14: Operating principle of in-plane switching.

the TFT array, so that the electric field is parallel to the glass plates. The lateral electric field causes the LC molecules to rotate parallel to the plates. Since the LC director always remains parallel to the glass plates, this leads to a fairly symmetric, very wide viewing cone. The dielectric anisotropy $\Delta\varepsilon$ of the LC fluid is usually positive. For $\Delta\varepsilon > 0$, the LC molecules line up parallel to the lateral electric field.

In Fig. 6.14 the basic operation of the IPS mode is shown. The alignment layers on both surfaces are rubbed under an angle of 45 degrees with the electrodes on the active matrix substrate. Linearly polarized light enters the cell parallel to the LC director. In the absence of an electric field, the orientation of the LC director is uniform throughout the cell and light emerges without a change in polarization. The exit polarizer is oriented perpendicular to the first polarizer and therefore blocks the light. When a voltage is applied between the electrodes in the ON state, the LC molecules line up to the field and the angle between the LC director and the polarization direction becomes gradually 45 degrees with increasing voltage.

The transmission T of this normally black IPS cell with crossed polarizers can then be described by the following equation:

$$T = T_0 \sin^2(2\theta(V)) \cdot \sin^2\left(\frac{\pi\Delta n.d}{\lambda}\right),\tag{6.2}$$

where T_0 is a constant depending on the polarizer transmittance and aperture ratio of the cell and $\Theta(V)$ is the angle between the LC optical axis and polarization angle of the incident light, which can be varied with the applied voltage V. Δn is the LC optical anisotropy, d is the cell gap, and λ is the wavelength. The cell gap is chosen so that $\Delta n.d \approx 0.3$ μm. This means that the last term in Eq. 6.2 is close to unity in the visible spectrum.

At $V = 0$, the LC optical axis is parallel to the incident polarization angle, so that $\Theta = 0°$ and $T = 0$. At high voltage (~7 V) most of the LC molecules are lined up parallel to the electric field (when $\Delta\varepsilon > 0$) so that $\Theta = 45°$ and $T = T_0$. Like the TN cell, the IPS cell is operated with an AC square wave voltage without the DC component.

The threshold voltage V_{th} for the IPS mode is

$$V_{th} = \pi\left(\frac{l}{d}\right)\sqrt{\frac{K_{22}}{\varepsilon_0 \Delta\varepsilon}},\tag{6.3}$$

where l is the spacing between the lateral electrodes across which the electric field is applied, K_{22} is the LC elastic constant for twist deformation, ε_0 is the permittivity in vacuum, and $\Delta\varepsilon$ is the dielectric anisotropy. V_{th} is the voltage at which the LC molecules in the center of the cell commence to rotate. It depends on l/d and is about 2–3 V. To reduce the threshold voltage and operating voltage of the IPS mode, special LC fluids have been developed with a large $\Delta\varepsilon$.

An example of the electro-optical response of an IPS cell for different wavelengths, corresponding to Eq. 6.2 is shown in Fig. 6.15.

Figure 6.16 shows some examples of pixel layouts for an IPS TFT LCD. The lateral electric field is applied between the drain electrodes, on which the data voltage is stored, and the counter electrodes, which are held at a constant bias. The pixel storage capacitor is formed by the overlap area of the counter electrode and the drain electrode. An ITO pixel electrode is not needed in this design. With both pixel electrodes on the TFT array, there is also no longer a need for a common ITO electrode on the color filters. However, to prevent sensitivity to external fields, a grounded ITO layer is often added as a shield between the front polarizer and the color plate. Since the pixel electrodes are opaque metal layers, they obscure the pixel opening significantly as compared to a TN TFT pixel. The pixel opening in IPS TFT LCDs is typically less than 50%.

Equation (6.3) shows that the threshold voltage of the IPS mode depends on both electrode spacing and the LC cell gap. In the TN cell, on the other hand, the electrode spacing is the same as the cell gap so that the threshold is independent, to a first approximation, of the cell geometry (see also Chapter 1, Sec. 1.4.). More accurate control of the cell gap in manufacturing is therefore needed for the IPS mode than for the TN mode.

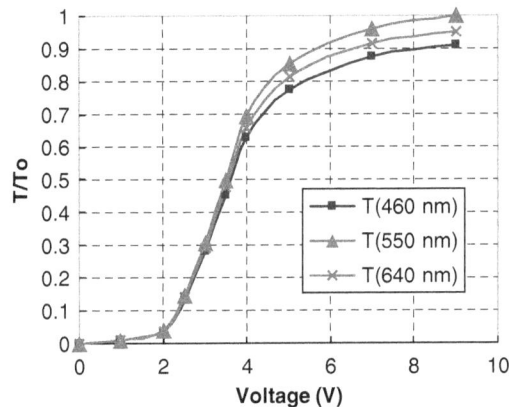

Figure 6.15: Transmittance-voltage curve for a typical IPS cell at different wavelengths.

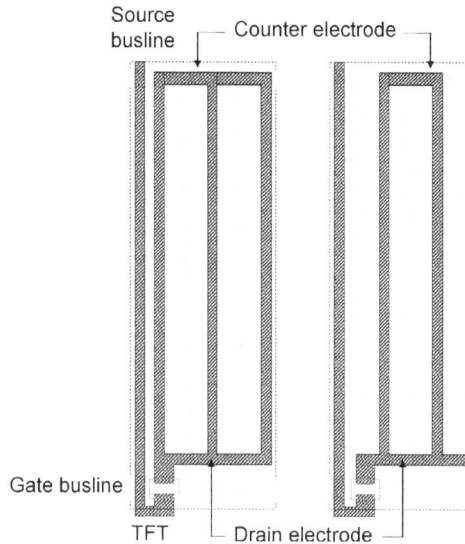

Figure 6.16: Examples of pixel layouts for an IPS cell.

Small electrode spacings give a lower operating voltage in the IPS mode. There is, however, a trade-off between operating voltage and aperture ratio. For a smaller electrode spacing, a larger fraction of the pixel area is occupied by the opaque metal electrodes. Aperture ratio is generally much lower than in the regular TN mode with or without compensation films. This is the price to pay for the viewing angle improvement. To maximize the aperture ratio, the electrode spacing has been increased from $l/d \approx 3$ to $l/d \approx 7$ in conjunction with the development of 15-V data driver ICs, which can apply +7 and −7 V on the pixel in the dot inversion mode.

The IPS mode was originally commercialized by Hitachi [6] and has been implemented by a number of companies in monitors and LCD televisions. Of all LC modes, it arguably maintains the best image quality over viewing angles. Not only is the contrast ratio larger than 10× up to 170-degree viewing cones, but the color shift and gamma shift over viewing angle are also minimal.

Drawbacks of the IPS mode are a somewhat lower peak contrast ratio at normal angles and a lower aperture ratio to accommodate the multiple electrodes on the pixel. The early IPS LCDs also had slow response times, exceeding 50 msec. With the development of better LC fluids specifically for IPS and with optimized cell structures, the response times have been reduced to be acceptable for television and contrast ratio has been improved to more than 600:1.

A number of variations and improvements on IPS, such as Super-IPS® introduced by Hitachi and True Wide IPS® introduced by LG Philips LCD, have pushed the state of the art even farther. With the original IPS pixel designs with the rectangular electrode configuration of Fig. 6.16, some asymmetry in the viewing cone occurred. In Super-IPS® LCDs the lateral pixel electrodes are in a chevron pattern, as shown schematically in Fig. 6.17.

Both the entrance polarizer axis and the initial LC molecule orientation at V = 0 are vertical. There is a difference in the electric field direction in different sections of the pixel, so that the LC molecules rotate in opposite directions. As a result, there is more than one domain in each subpixel. This tends to further widen the viewing angle and improve symmetry of the viewing cone. It also reduces the color shift over viewing angle [7]. In Fig. 6.17 the angle between the polarizer and the lateral electric field is about 60 to 70 degrees. In this case, the angle Θ between the entrance polarizer and the LC director can exceed 45 degrees so that the transmittance-voltage characteristics show a peak at $\Theta(V)$ = $\pi/4$ and then decline (Fig. 6.18), in accordance with Eq. 6.2.

The rise and fall response times in the IPS mode are given by

$$\tau_r = \frac{\gamma_1}{\varepsilon_0 \Delta\varepsilon E^2 - \dfrac{\pi^2 K_{22}}{d^2}}, \tag{6.4}$$

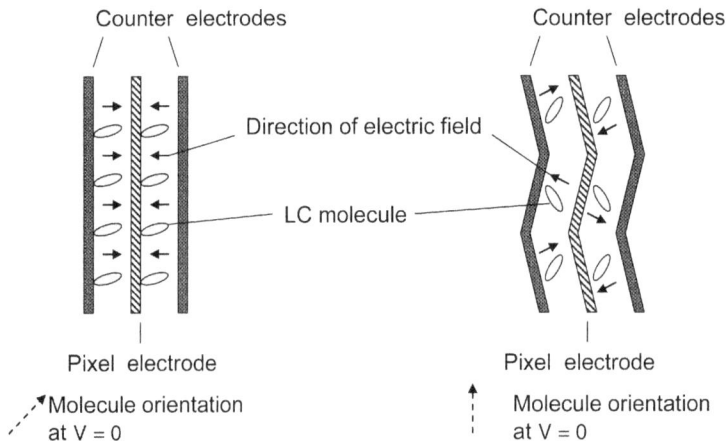

Figure 6.17: Electrode configuration of a conventional IPS (left) and a Super IPS (right) pixels.

$$\tau_f = \frac{\gamma_1 d^2}{\pi^2 K_{22}},\qquad(6.5)$$

where E is the electric field applied between the lateral electrodes. As in the TN mode, the response times depend strongly on cell gap d. Reducing d improves video performance but increases the threshold voltage of the IPS mode (Eq. 6.3). Since maintaining uniformity in manufacturing becomes increasingly more difficult for thinner cell gaps, a practical value is around 4 μm. With the development of LC fluids with high $\Delta\varepsilon$ specifically for the IPS mode, good video performance has been achieved.

A variation on the IPS mode is the fringe-field-switching (FFS) mode introduced by BOE Hydis. It has excellent viewing angle, comparable to IPS, and minimizes one of the drawbacks of IPS: the limited transmittance. In the FFS mode a higher aperture ratio is obtained by the use of a continuous common ITO electrode separated from ITO grid electrodes by a dielectric. Both electrodes are on the active matrix array and are transparent, so that the pixel opening can be significantly larger than in most IPS TFT LCDs, where both electrodes are opaque.

6.4.3   Vertical Alignment

Another successful approach to improve viewing angle is the vertical alignment (VA) mode (Fig. 6.19). In the VA mode the nematic LC molecules are aligned perpendicular to the glass plates in the absence of an electric field. This requires a different type of alignment layer. The rubbing process is eliminated in the VA mode, which translates into a yield advantage.

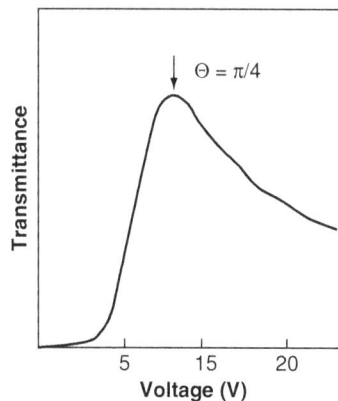

Figure 6.18: IPS T-V curve when Θ can exceed $\pi/4$.

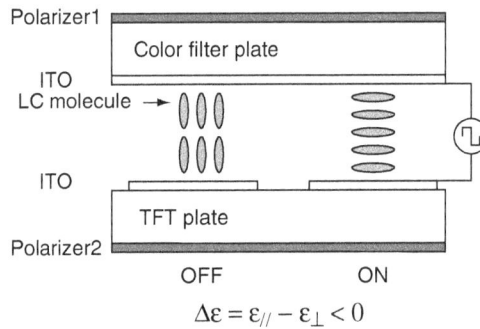

$$\Delta\varepsilon = \varepsilon_{//} - \varepsilon_{\perp} < 0$$

Figure 6.19: Operating principle of the vertical alignment (VA) mode.

LC fluids for the VA mode are specially formulated to have a negative dielectric anisotropy $\Delta\varepsilon$ of -3 to -4. These LC fluids have lateral polar substituents that induce a dipole moment perpendicular to the long axis (the director) of the molecule, hence the negative dielectric anisotropy.

This means that when a voltage is applied between the transparent electrodes on the two plates, the long axis (the director) of the LC molecules rotates to become oriented parallel to the plates. The operating principle of the VA mode is illustrated in Fig. 6.19.

In the basic VA mode the molecules can start tilting with their director in any plane perpendicular to the glass substrates. For situations with partial tilt, this would lead to domains with varying retardation under oblique angles and therefore varying, uncontrolled contrast.

In order to obtain a more symmetrical viewing angle, the pixel is subdivided into domains, in each of which rotation starts with well-defined different initial tilt. This is called the multi-domain vertical alignment (MVA) mode (see Fig. 6.20). The multiple domains are obtained by adding protrusions on both substrates. They cause the pixel to have areas of different preferential tilt direction in the cell when an electric field is applied between the two substrates.

The cell operates with crossed polarizers in the normally black mode. Without applied field, most LC molecules line up perpendicular to the substrate surfaces so that the polarization is not changed. The exit polarizer blocks the light. When a moderate field is applied, the LC molecules tilt preferentially in a direction controlled by the protrusions so that the polarization is partially altered to obtain a mid-gray level. At high field, the center molecules line up parallel to the surfaces as a result of the negative dielectric anisotropy. The LC fluid now rotates the polarization direction depending on the angle of the entrance polarizer and the LC director. The two or four domains have the effect of making the viewing angle symmetric in two or four directions.

Figure 6.21 shows a pixel layout of a protrusion design to obtain four domains per pixel for the MVA mode. A zigzag grid of protrusions at 90 degrees on the active and color

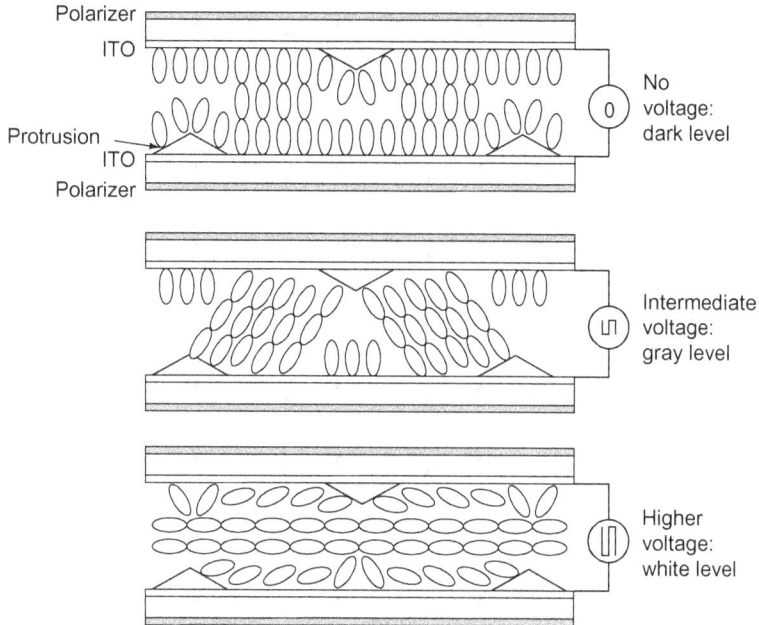

Figure 6.20: Multi-domain vertical alignment (MVA) using protrusions on the two substrates.

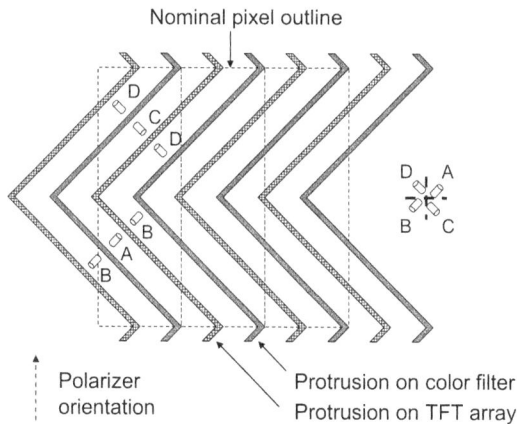

Figure 6.21: Protrusion design for a four-domain MVA LCD [8].

plates is used. The entrance polarizer is vertical along the column lines so that the LC director in each of the four domains will be oriented at +/–45 degrees with the polarizer orientation in the presence of an electric field. The protrusions block the light and therefore need to be kept small to minimize impact on the aperture ratio. The aperture ratio, however, will be smaller than in a regular TN mode with or without compensation films. In Fig. 6.22 a transmittance-voltage curve for the MVA mode is shown.

The MVA mode is popular in desktop monitors and LCD televisions, and combines a wide viewing angle with a fast response time. It was originally introduced by Fujitsu [8] and has been adopted by a number of manufacturers.

Variations and improvements, called patterned vertical alignment (PVA® and super-PVA®), and premium-MVA® mode, have been introduced by Samsung and AU Optronics, respectively. Some improvements have protrusions on only one substrate and aperture ratios as high as 67% have been achieved. Of particular interest is Samsung's PVA mode, which eliminates the protrusions altogether and instead uses patterned ITO areas on both the TFT and color substrate to obtain multiple domains, as shown in Fig. 6.23. An extra ITO patterning and etching step on the color plate is needed for the PVA process. As in the MVA mode, a four-domain design is common with zigzag-shaped slits in the ITO layers. To maximize aperture ratio, the data buslines can also be designed in a zigzag pattern [9].

In the dark state of the PVA mode (V = 0), the LC molecules are perfectly lined up perpendicular to the glass plates. As a result, there is no residual retardation and the contrast ratio can be virtually equal to the extinction ratio of the polarizers. With the optimized

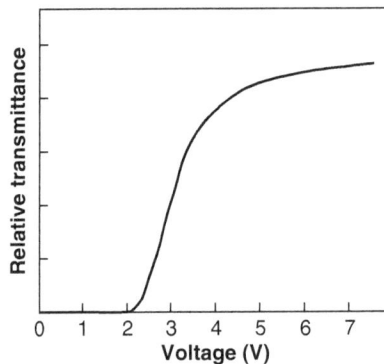

Figure 6.22: Typical transmittance curve for an MVA LCD.

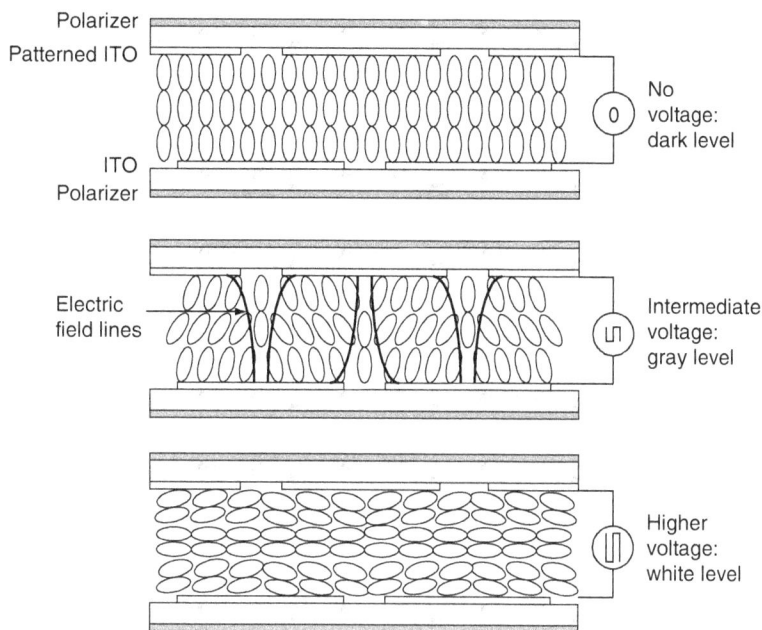

Figure 6.23: Patterned vertical alignment (PVA) mode.

super-PVA mode [9], a contrast ratio larger than 1000 at normal viewing angles has been achieved (Fig. 6.24). A contrast ratio exceeding 1000:1 is particularly important for LCD television where any light leakage in low ambient lighting conditions is objectionable.

In the protrusion-based MVA mode it is more difficult to achieve a contrast ratio of 1000:1 because the small tilt near the protrusions causes some residual retardation and light leakage in the dark state (V = 0). IPS-mode LCDs have a lower peak contrast ratio as well, but usually have an even better contrast ratio off-axis than is shown in Fig. 6.24. As compared to the original contrast versus viewing angle curve of a standard TN cell shown in the previous chapter (Fig. 5.6), the MVA, PVA, and IPS modes have all demonstrated tremendous progress in image quality.

MVA-mode LCDs use a special alignment layer to cause the vertical orientation of the LC molecules at the substrate surfaces. Unlike the TN and IPS modes, this orientation layer does not need a rubbing process, which is considered a yield advantage in manufacturing for the MVA and PVA modes.

To compensate for the off-axis retardation in the zero field dark state, MVA-mode AMLCDs do, however, usually require retardation films added outside the glass assembly.

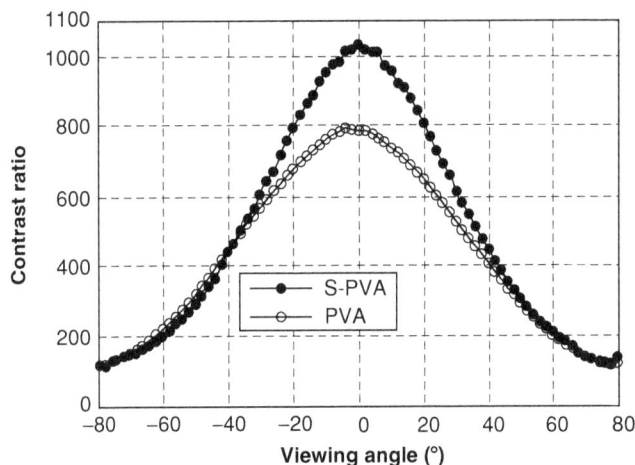

Figure 6.24: Contrast ratio versus viewing angle for PVA and S-PVA mode (reprinted with permission from the Society for Information Display).

A variation on the VA mode is the advanced super view or axially symmetric vertical alignment (ASV®) mode used by Sharp in their LCD televisions. In the ASV method (Fig. 6.25), the domains have random directions leading to a completely symmetric viewing cone.

6.4.4 A Comparison and Other Viewing Angle Improvement Methods

The viewing angle behavior of LC cells can be quantitatively expressed as contrast ratio versus left/right and upper/lower viewing angles, as shown in Figs. 5.6 and 5.8 in Chapter 5. These plots show the performance only at azimuth angles of 0, 90, 180, and 270 degrees. To include all azimuth angles, it is customary to plot the viewing angle behavior in contrast contour plots with azimuth angle ϕ and polar angle θ as parameters, as defined in Fig. 6.26.

These two angles are used to generate the contrast contour plots shown in Fig. 6.27 for unmodified TN-, compensated TN-, IPS-, and MVA-mode cells. The iso-contrast curves enclose areas with higher contrast. The plots are representative for measured results with the caveat that improvements are still regularly reported and that there is some variation from one manufacturer to the next. For both the MVA mode and the IPS mode, a contrast ratio exceeding 10:1 has been achieved for polar angles up to 80 degrees at all azimuth angles.

There are a few other, less popular methods to improve viewing angle, notably the multigap normally black TN mode and the TN mode with collimator-diffuser combination.

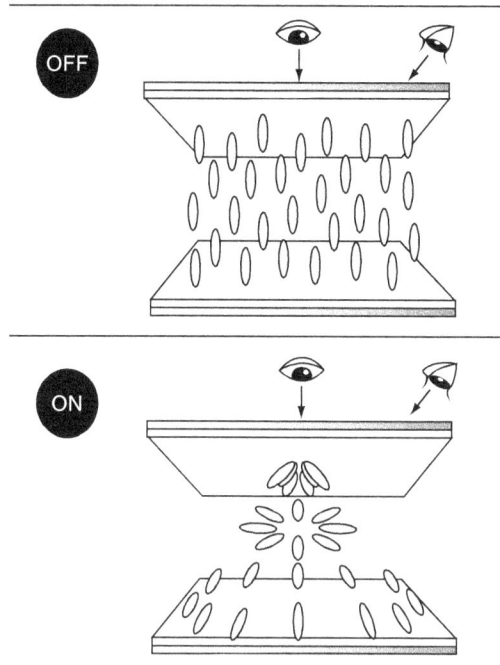

Figure 6.25: Operating principle of the ASV mode.

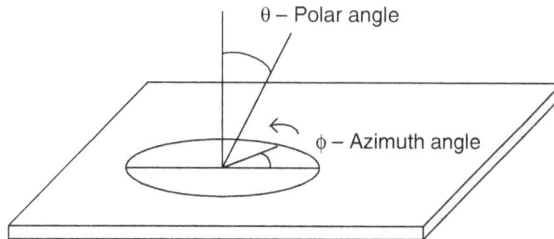

Figure 6.26: Definition of polar and azimuth angles for measuring and plotting contrast contours.

The normally black (NB) TN mode has better contrast ratio at off-angle viewing than the normally white mode, but a lower peak contrast ratio with a white backlight. The contrast ratio varies with wavelength and LC cell gap according to the Gooch–Tarry theory (Fig. 6.28), as was described in Chapter 5, Sec. 5.2.

The NB TN LCD can be designed with different cell gaps for each color so that the cell operates in the first minimum for red, green, and blue light. This is done by tailoring the cell gaps according to the following equations:

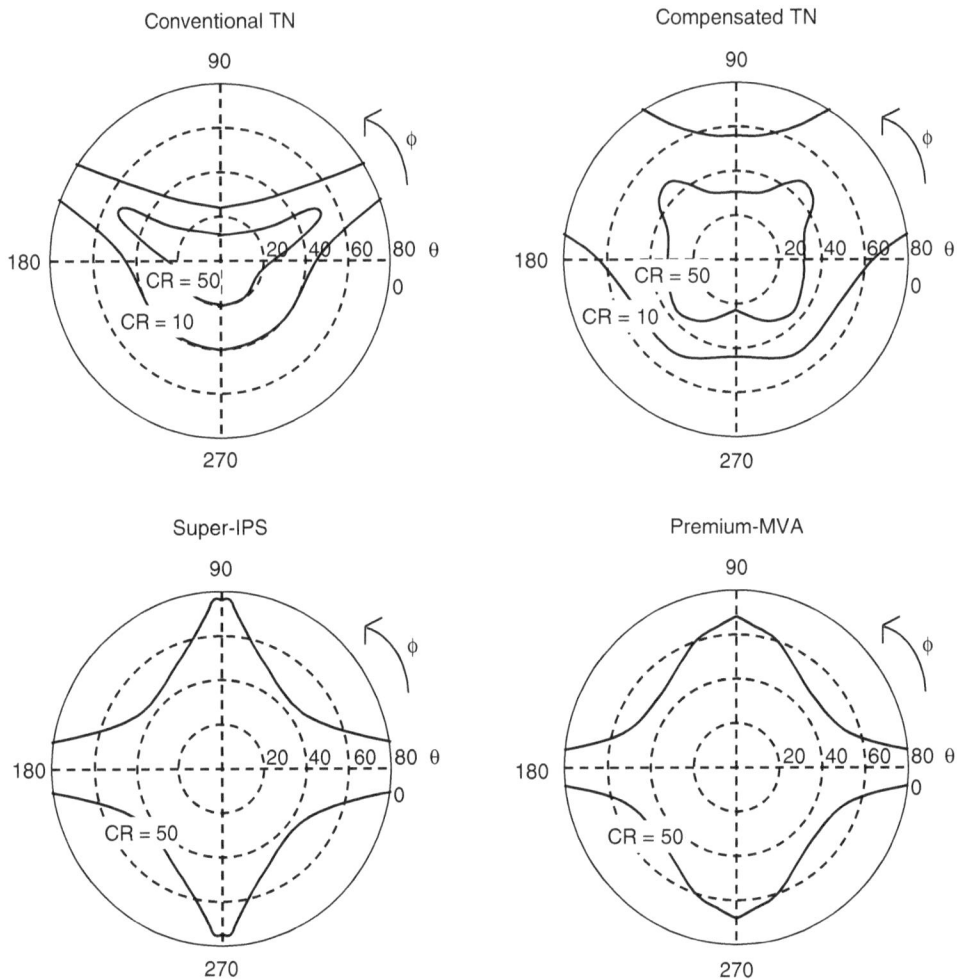

Figure 6.27: Contrast contour plots for four different LC modes.

$$d_R = \frac{\sqrt{3}}{2}\frac{\lambda_R}{\Delta n},$$ (6.7)

$$d_G = \frac{\sqrt{3}}{2}\frac{\lambda_G}{\Delta n}, \text{and}$$ (6.8)

$$d_B = \frac{\sqrt{3}}{2}\frac{\lambda_B}{\Delta n};$$ (6.9)

where d_R, d_G and d_B are the LC cell gaps for the red, green, and blue subpixels and λ_R, λ_G and λ_B are the peak wavelengths for the three primary colors. The color filter thicknesses are adjusted to obtain the different cell gaps, as shown in the inset of Fig. 6.28. A high peak contrast ratio and relatively wide viewing angle can be simultaneously achieved this way. However, the multiple cell gaps are very difficult to control accurately and consistently in manufacturing, and this mode has therefore been used only in high-end avionic cockpit displays but not in mainstream displays.

The collimator-diffuser approach (Fig. 6.29) avoids altogether the problem with off-angle viewing in the TN mode by collimating the light so that it passes through the LC cell

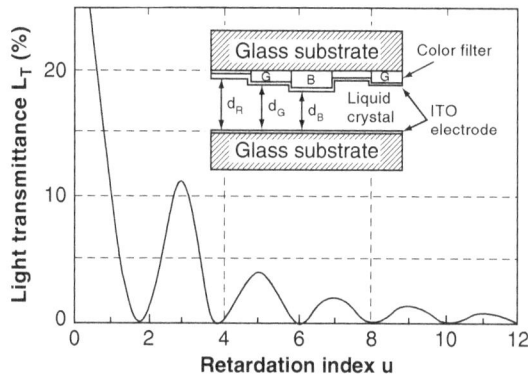

Figure 6.28: OFF state transmission in normally black TN mode with multi-gap configuration in inset.

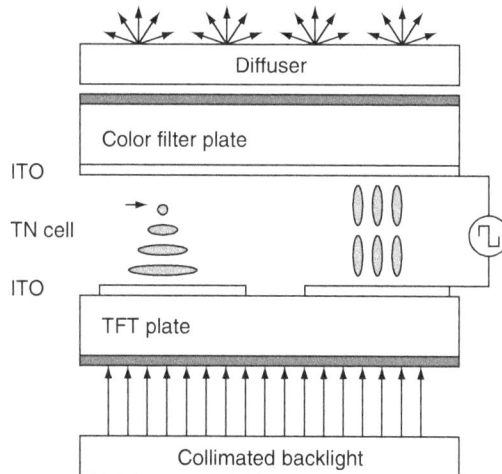

Figure 6.29: Operating principle of the collimator-diffuser mode.

mostly perpendicular to the glass plates with good contrast. A diffuser after the exit polarizer scatters the light. Its drawback is increased power consumption in the backlight and light loss in the front diffuser.

In Table 6.2 the major three viewing angle enhancement methods are compared with each other and with the conventional TN mode in terms of image quality, power consumption, and manufacturability. All three methods give much better image quality over angles than the conventional, uncompensated TN mode. Response times are not listed in the table because all three basic modes. TN, IPS, and MVA, have made similar progress toward video performance.

There is still active ongoing research to improve the different modes for televisions and monitors in terms of response times, color shift over angles and gray scale, and aperture ratio. The improvements and refinements are labeled with superlatives, some mentioned before, such as Super-IPS®, Advanced Super-IPS®, True Wide IPS®, super PVA® and premium MVA®.

6.5   Enhancement of Video Performance

When conventional LCD desktop monitors and notebook panels are used to display video from a DVD, for example, the image quality is inferior to that on a CRT. Although the contrast ratio for static images on such displays can exceed 500:1, for moving images degradation in the dynamic contrast ratio is seen and the edges of moving objects blur. This makes the viewing experience less than acceptable. Faster LC fluids have been developed with a lower rotational viscosity and the cell gap can be reduced to about 4 μm to improve response time (see Eqs. 5.11 and 5.12 in Chapter 5). These measures can lower the transition time from full ON to full OFF and vice versa to 8 msec for TN and MVA modes.

Table 6.2: Comparison of different viewing angle improvement modes

	Conventional TN	Compensated TN	IPS	VA
Viewing angle	-	x	0	Δ
Peak CR	Δ	Δ	Δ	0
Gamma shift over angle	-	x	0	Δ
Color shift over angle	-	x	0	Δ
Power consumption	0	0	x	Δ
Manufacturability	0	0	Δ	Δ

0, very good; Δ, good; x, less; -, poor.

However, this by itself does not reduce smearing of moving video images sufficiently for application of LCDs in televisions. In video imagery the majority of transitions are between intermediate gray levels, which have longer response times, according to Eq. 5.11. The response time between different gray levels can exceed 50–80 msec in a conventional TN cell.

Another factor influencing motion portrayal is the hold-type character of an LCD (i.e., the luminance is continuous during each frame time). A data voltage is applied to each pixel once per frame time and maintained during the entire frame until the next refresh. This sample-and-hold mode of operation cannot eliminate an afterimage on the retina and causes blurring of fast-moving objects on the LCD. The problem is exacerbated when the display is large and/or viewed from close-up. This is contrary to the impulse-type character of CRTs, in which each pixel emits light for only a fraction of the frame time.

In the following two sections, two solutions for these basic problems in LCDs are described:

1. Response time compensation by overdrive techniques, and

2. Emulation of impulse-type operation in an LCD.

6.5.1 Response Time Compensation

In Sec. 5.5 in the last chapter, the response times for the TN mode were given in Eqs. 5.11 and 5.12. These are the times for transitions from full ON to full OFF and vice versa. The rise time is voltage-dependent, so for transitions between intermediate gray levels the response times can exceed several frame times and are not short enough for high-quality video. The resulting image lag and motion blur is unacceptable for LCD televisions. In addition to rise and fall times of the LC cell in excess of a frame time, a contributing factor is the slow variation of the LC capacitance with voltage. When the data voltage on the pixel changes significantly, it can take more than a frame time for the LC capacitance to stabilize at its new value. Since a constant charge is loaded onto each pixel during active matrix addressing, the voltage may change during the frame time from V_{start} to V_{end} according to the equation

$$V_{end} = \frac{C_{start}}{C_{end}} V_{start},$$

(6.10)

where C_{start} and C_{end} are the LC capacitance at the beginning and end of the frame time, respectively.

The adverse effects of both slow response times and dynamic LC capacitance can be countered by response time compensation [10]. To improve video performance, overshoot data voltages are used so that the target luminance is reached after one frame time. The principle of this response time compensation is illustrated in Fig. 6.30. The final gray level *i* cannot be obtained within one frame time by applying data level *i*. Instead, an overshoot data signal or boost level *k* is applied to the pixel and is stored on the pixel for one frame time. The overshoot level causes the display luminance to settle at the new gray level *i* within one frame time. The overdrive method requires signal processing and empirically optimized lookup tables for level-to-level transitions.

An example of the electronics required to implement response time compensation is shown in Fig. 6.31. One frame of incoming video signals is stored in a first in, first out (FIFO) frame buffer. The video data for each pixel in this frame is then compared with the video data for the next frame. Depending on the starting and ending gray level, an overshoot or undershoot voltage is added to the steady-state data signal for a particular gray level. The signal boost for each gray level transition is derived from a lookup table,

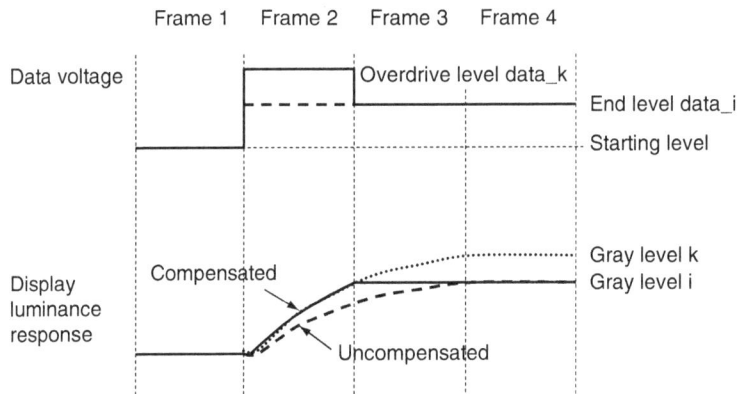

Figure 6.30: Overdrive data levels applied to speed up optical response time of display luminance between gray levels.

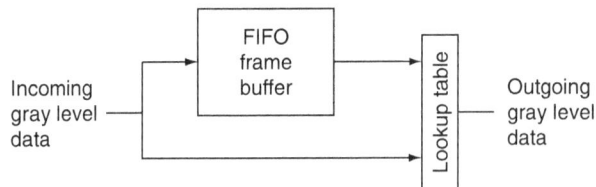

Figure 6.31: Basic block diagram for response time compensation.

which is optimized for the particular LC mode and cell structure used (TN, IPS, or MVA). The boost signal is usually inserted during one frame only, after which the steady-state gray level is applied (assuming no change in gray level in the subsequent frame).

The overdrive functionality with the lookup table and the comparison circuit can be integrated into the timing controller of the display or in separate added circuitry. The frame buffer memory is typically external in an SDRAM chip.

The plots in Fig. 6.32 demonstrate typical improvements in response times achieved by the overdrive method. When combined with faster LC fluids, response times of 8 msec have been obtained for all gray level transitions.

6.5.2   Emulation of an Impulse-Type Display

Even when the response times could be reduced to zero, the video performance of LCDs would still be inferior to that of CRTs. CRTs are impulse-type displays in which brief excitations of the phosphor result in brief repeating light pulses, which are integrated by the human eye and perceived as continuous while they are actually intermittent (Fig. 6.33). On the CRT, the edge of a moving object therefore remains sharp.

In an LCD (a hold-type display), the luminance is continuous. The eye traces a fast-moving object on the LCD screen as indicated by the arrow in Fig. 6.33. When integrating over the frame time, the average luminance leads to a blurry edge on the object.

One method to improve video performance is increasing the frame rate to, for example, 120 Hz. When all the gray-to-gray level transition times are also reduced to 8 msec, a faster frame rate of 120 Hz gives a clearly visible improvement. However, this does not solve the problem entirely and it also complicates the electronics and requires larger TFTs to charge up the pixels in half the time.

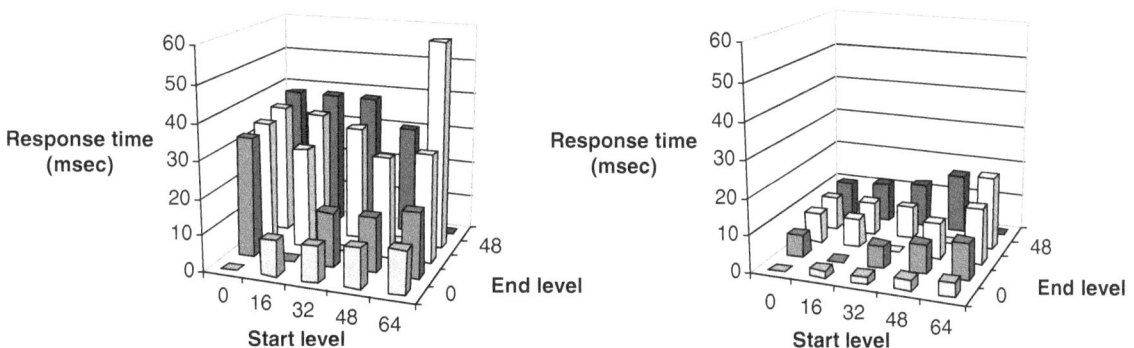

Figure 6.32: Example of uncompensated (left) and compensated (right) gray level response times.

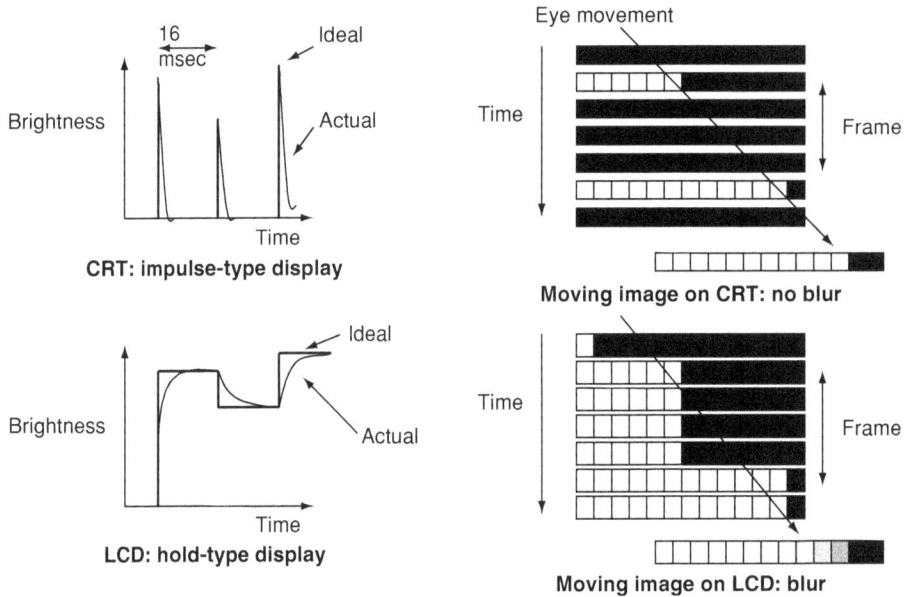

Figure 6.33: Brightness versus time for and perception of moving objects on a CRT and an LCD.

Another way to address the problem is to switch the backlight ON and OFF to get the impulse-type operation. This can be done be flashing the entire backlight at once after writing the data information on the displays, as shown in Fig. 6.34A. The LC transmittance has to be stabilized before the backlight turns ON. Part of the frame time is used for writing data and the second part for illumination. This causes a problem because the allowable LC response time for pixels at the bottom of the display will be shorter than for the top, causing moving objects with blurred edges at the bottom of the display for practical finite response times.

A better solution is a scanning backlight with, for example, six stick lamps and the timing shown in Fig. 6.34B. The physical configuration is depicted in Fig. 6.35. In this case the entire frame time can be used for writing data and the allowable LC pixel response time is the same at the top and the bottom of the display. An LC response time for all gray-to-gray levels of about two-thirds or one-half the frame time is sufficient to prevent moving objects with blurred edges.

The use of this scanning backlight emulates an impulse-type display and virtually eliminates blur of moving objects, as illustrated in Fig. 6.36. Motion on LCDs with scanning backlight appears more like a movie. Film projectors also darken the screen for about 50% of the time, during the mechanical advance to the next film frame. In the projector this is done by a shutter behind the lens.

(A)

(B)

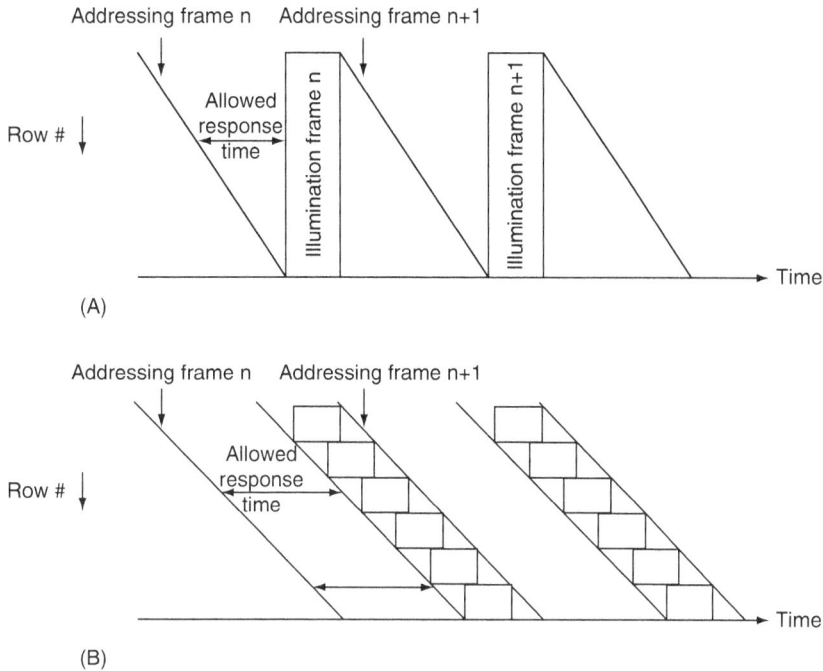

Figure 6.34: Synchronization of blinking backlight (A) and scanning backlight (with six lamps) (B) to vertical scanning of the LCD.

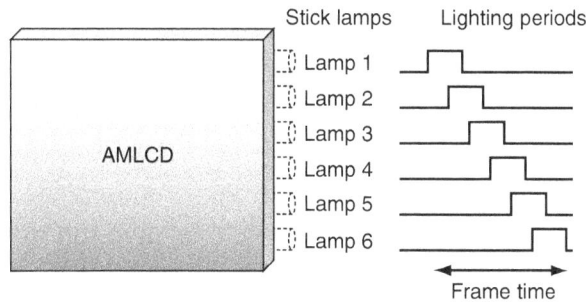

Figure 6.35: Operation of AMLCD with scanning backlight to improve video performance.

When one lamp in the scanning backlight is ON at a time and its brightness is not increased, an obvious drawback is that the average brightness is decreased by a factor of six. Since this is unacceptable, the lighting periods of the individual lamps will overlap in practical scanning backlights. There will be a trade-off between brightness and motion blur.

171

Direction of motion

Blurred edge of moving objects
with conventional backlight

Sharper edge of moving objects
with scrolling backlight

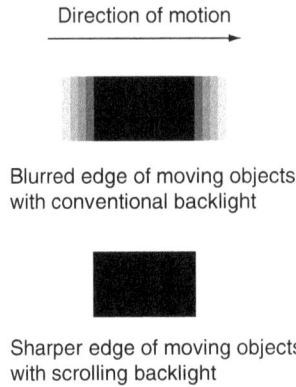

Figure 6.36: Qualitative improvement in motion portrayal with impulse-type (scanning or scrolling) backlight.

The scanning or scrolling backlight improves not only sharpness, but also the contrast and color purity of moving images. As explained in Chapter 4, Sec. 4.5, hot cathode fluorescent lamps (HCFLs) are a good match with scanning backlights. They can operate at higher currents to obtain very high luminance and can also be easily dimmed for intermittent operation. In the most advanced scanning backlights, both duty cycle and dimming level are dynamically controlled based on the video content and the ambient lighting conditions. With the gamma curve of the LCD also being dynamically adjustable depending on image content and ambient lighting, superior display performance is achieved with a sophisticated, correlated control of backlight and LCD electronics.

LEDs are compatible with scanning backlights as well. The peak current in the LEDs can be increased for shorter duty ratios so that the average luminance can be maintained quite well. LED backlights are also relatively easy to operate intermittently with current pulses at low voltage and will be attractive for scanning backlights, when their cost can be further reduced.

Another method to emulate an impulse-type display is to insert black data during part of the frame. To avoid flicker this needs to be done during part of the 16-msec frame and therefore requires very fast LC response times.

6.6 Large Size

With the advent of Generation 5, 6, and 7 production lines for a-Si TFT LCDs, it has become possible to build very large displays with a size approaching that of the largest plasma displays. Before LCDs larger than 40 in. or with resolution higher than SXGA became possible, several technical hurdles had to be overcome. One is, of course, manufacturing yield, discussed in Chapter 3, Sec. 3.9. Another is the line resistance of the select and data buslines. When the signal propagation delays on the buslines becomes a significant fraction of the line select time, LCD performance can deteriorate.

This is the case also for 20- to 25-in. displays with very high resolution in the UXGA to QUXGA range, developed for graphics and medical imaging applications. They have a very short line select time, as shown earlier in Table 5.3.

Table 6.3: Resistivities of metal films used as buslines in LCDs

Metal	Resistivity ($\mu\Omega$cm)
Ta	80–200
Cr	20
Mo	13
AlNd	7
Al	4
Cu	2.5

Long buslines or short select line times both require the use of low-resistance gate lines to minimize the RC propagation delays. Figure 6.37 shows how the select pulse is distorted along the select line of a large or high-resolution display. The load on each pixel can be represented by a resistance R/N and capacitance C/N, where N is the number of pixels on a row. A distributed network of these capacitors and resistors represents the entire row. Both the rise time and the decay time of the gate pulse are increased at the end of the row line.

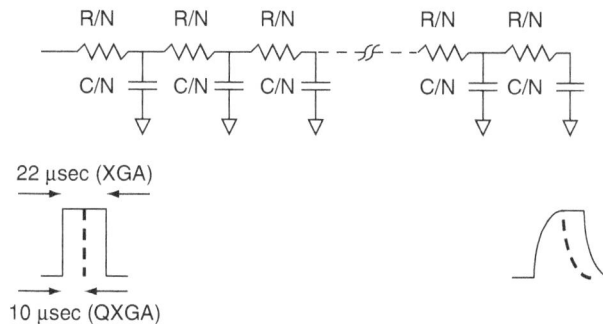

Figure 6.37: Equivalent circuit for gate busline with resistance R, capacitance C, and N pixels with example of gate pulse distortion along rows in XGA and QXGA cases.

With the aid of circuit simulations the effect of propagation delays on the LC pixel voltage can be predicted. The pixel charging is faster for the negative charging cycle than for the positive cycle, as a result of the higher gate voltage relative to the negative data voltage than to the positive data voltage. The pixel may not charge up completely to the positive data voltage (Fig. 6.38).

After the negative charging cycle, the pixel voltage shift from the parasitic gate-drain capacitance further reduces the voltage on the ITO pixel electrode. Because of the slow decay time of the gate pulse, the TFTs at the end of the row are not sufficiently turned OFF before the data voltage changes for the next pixel on the column. Then, there is a potential for partial recharging of the pixel to the next pixel's data voltage, as shown in Fig. 6.38.

Incorrect pixel charging leads to gray scale errors and/or residual DC voltage on the LC (and therefore image retention).

Conventional select line metals such as Ta, Cr, and Mo have a relatively high resistivity of more than 12 μΩcm (see Table 6.3 and Sec. 3.3 in Chapter 3). A high gate line resistance can show up in the display as a brightness gradient from the driven side to the undriven side or as an increase in flicker and image retention in sections of the viewing area. The RC delay is most severe when the storage capacitor is connected to the gate

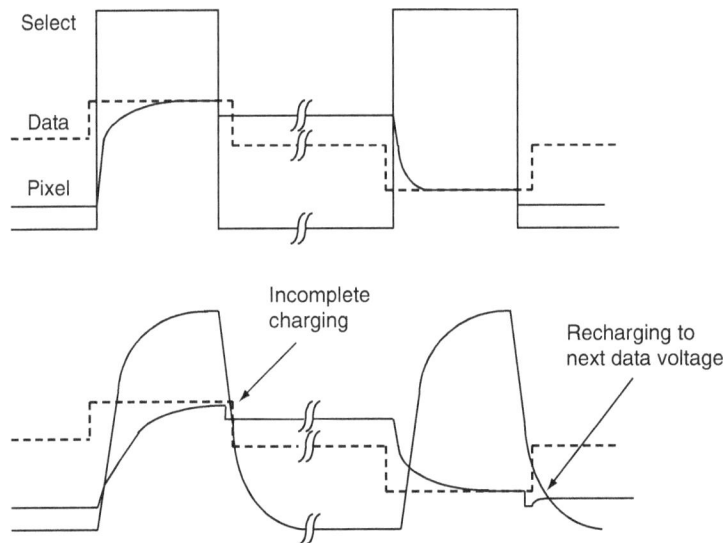

Figure 6.38: Pixel wave forms at beginning (top) and end (bottom) of gate line with RC delay.

line because it increases the capacitive load on the gate line. When the Cs-on-common pixel configuration (see Fig. 2.20 in Chapter 2) is used, this problem is somewhat alleviated. For very large displays there is also the option to drive the gate lines from both left and right side, so that the worst-case RC delay occurs in the center pixels. Twice the number of row drivers will be required. In this configuration, both R and C are reduced by a factor of two, resulting in a 4× reduction in the worst-case RC propagation delay. RC delays on the data buslines are also important, although not as critical for performance as RC delays on the gate lines.

Most large and high-resolution AMLCDs use Al as the busline material for both select and data lines and some use Cu (see Table 6.3).

Al and Cu have a much lower resistivity of around 4 and 2.5 $\mu\Omega$cm, respectively. Unfortunately, pure Al grows hillocks during high-temperature processing such as the PECVD step for the gate nitride and a-Si. This can cause shorts and yield problems. To prevent hillock formation, the Al layer can be anodized, encapsulated, or covered by a cap layer such as Mo or Ti (Fig. 6.39). Alternatively, an Al alloy with, for example, Nd can be used (Fig. 6.39). Since anodization and encapsulation require extra patterning steps, they are used less frequently than cap layers or Al alloys.

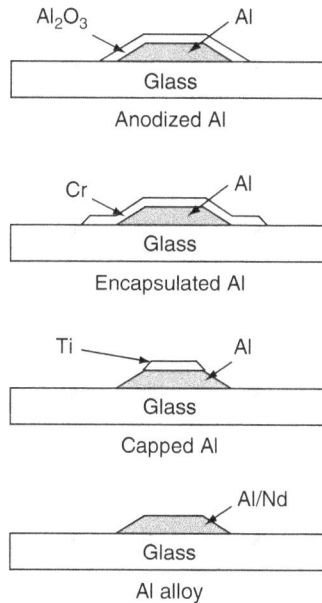

Figure 6.39: Methods to employ Al gate buslines.

Figure 6.40: Photograph of a 57-in. TFT LCD developed by Samsung Electronics.

The largest reported a-Si TFT LCDs are an 82-in. HDTV by Samsung Electronics, a 65-in. HDTV by Sharp and a 57-in. HDTV by Samsung Electronics [10]. The 57-in. display shown in Fig. 6.40 has a resolution of 1080×1920 pixels and utilizes dual-sided row drive to reduce both resistance and capacitance of the select lines by 2× for a reduction in RC delay of 4×. Viewing angle is enhanced by the PVA (patterned vertical alignment) mode. Response time is improved for television applications by overdrive.

The application of large TFT LCDs beyond 30 in. is mostly in enhanced-definition and high-definition television and in digital signage. In both markets they compete with plasma displays.

References

1. W. den Boer, J.Z.Z. Zhong, T. Gu, Y.H. Byun, and M. Friends, "High Aperture TFT LCD Using Polymer Interlevel Dielectric," *Proc. Eurodisplay '96*, pp.53–56 (1996).
2. J.Z.Z. Zhong, W. den Boer, Y.H. Byun, and S.V. Thomsen, "New AMLCD Structure—Color Filter Passivated a-Si TFT," *Proc. Asia Display '98*, pp. 113–116 (1998).

3. B.W. Lee, C. Park, S. Kim, T. Kim, Y. Yang, J. Oh, J. Choi, M. Hong, D. Sokong, and K. Chung, "TFT LCD with RGBW Color System," *SID 2003 Digest*, pp. 1212–1215 (2003).
4. Search the Internet for "3M Optical Films" (with the Vikuity™ trade name)
5. H. Mori and P.J. Bos, "Designing Concepts of the Discotic Negative Birefringence Compensation Films," *SID 1998 Digest*, pp. 830–833 (1998).
6. K. Kondo, K. Kinugawa, N. Konishi, and K. Kawakami, "Wide-Viewing-Angle Displays with In-Plane Switching Mode of Nematic LCs Addressed by 13.3-in. XGA TFTs," *SID 1996 Digest*, pp. 81–84 (1996).
7. Z. Tajima, "IPS Technology Trends," *Asia Display/IMID '04 Digest*, pp. 15–18 (2004).
8. A. Takeda, S. Kataoka, T. Sasaki, H. Chida, H. Tsuda, K. Ohmuro, T. Sasabayashi, Y. Koike, and K. Okamoto, "A Super-High Image Quality Multi-domain Vertical Alignment LCD by New Rubbing-Less Technology," *SID 1998 Digest*, pp. 1077–1080 (1998).
9. K.H. Kim, N.D. Kim, D.G. Kim, S.Y. Kim, J.H. Park, S.S. Seomun, B. Berkeley, and S.S. Kim, "A 57-in. Wide UXGA TFT-LCD for HDTV Application," *SID 2004 Digest*, pp. 106–109 (2004).
10. H. Okumura and H. Fujiwara, "A New Low-Image-Lag Drive Method for Large-Sized LCTVs," *SID Digest 1992*, pp. 601–604 (1992).

Special AMLCD Configurations

In this chapter, a number of specific AMLCD configurations are described that are different from the standard, stand-alone transmissive color TFT LCD.

Ultra-high-resolution monochrome monitors with more than 10-bit gray scale have been developed for medical imaging to replace X-ray film on light boxes.

Reflective LCDs for low-power portable applications are briefly addressed along with their pros and cons as compared to transflective LCDs, which use a combination of the reflective and transmissive operational modes.

In field-sequential color LCDs the color filters are eliminated and temporal rather than spatial color mixing is employed with the potential for higher resolution and lower power.

In this Chapter, emerging stereoscopic LCD technology will be introduced as well. Finally, various types of touch screens added to LCDs will be described. They facilitate a more interactive and intuitive user input.

7.1 Ultra-High-Resolution Monitors and Improved Gray Scale

The highest resolution display to date is a 22-in. W-QUXGA display with 9.2 million pixels, as demonstrated by IBM and produced in low volumes [1]. It uses a multi-domain IPS mode for wide viewing angle and needs several DVI connectors to supply the video signal.

One important application of high-resolution TFT LCDs in the 2–5 Mpixel range is in medical imaging. With the advent of digital X-ray detectors (also using a-Si TFT arrays and to be described in Chapter 9), X-ray films are being replaced with digital images that need to be studied on high-quality monitors. Other modalities such as MRI, ultrasound, and computed tomography also need high-quality displays, although generally not as high-resolution as X-ray imaging.

The monochrome image on the LCD needs to have the same or better image quality than X-ray film on a light box, which means up to 700 nits brightness and more than 10-bits of gray scale. Examples are the C2, C3, and C5i medical-imaging monitors of Planar Systems, Inc. [2], shown in Fig. 7.1. To obtain a high-luminance, monochrome display, the color filters are left out while keeping three subpixels per pixel. The number of gray levels can be increased by applying gray level signals varying by one least significant bit to the individual subpixels, as shown in Fig. 7.2. This allows the continued use of standard 8–bit data driver ICs with only 256 gray levels to obtain 766 gray levels on the display. The other approach to increase the number of gray levels, described in Chapter 4, Sec. 4.1, is the frame rate control technique, also illustrated in Fig. 7.2. When using this temporal dithering technique over four frames, a display with 8-bit data drivers can achieve a 10-bit gray scale.

By combining spatial and temporal modulation in the medical imaging displays, the number of gray levels has been further increased to 3061. X-ray imaging monitors are a demanding application for contrast and luminance uniformity because radiologists are used to very high-quality images on X-ray film. Any non-uniformity or *mura* on the LCD could be misinterpreted as a pathology.

The 2-, 3-, and 5-Mpixel LCDs are used for diagnostics in radiography and mammography, and employ the multi-domain IPS mode for optimum gray level consistency over angles.

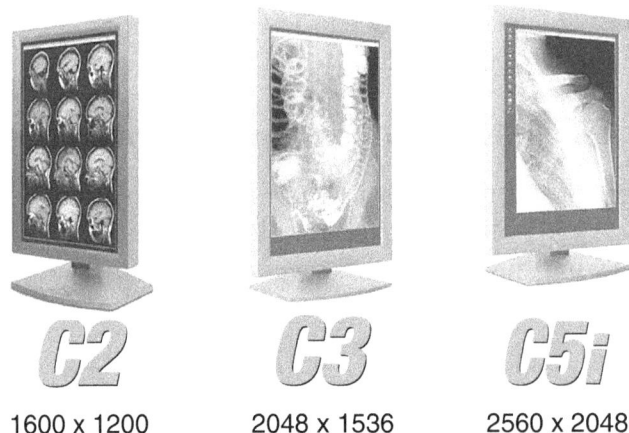

C2	C3	C5i
1600 x 1200	2048 x 1536	2560 x 2048

Figure 7.1: High-resolution monochrome medical imaging displays with 2, 3, and 5 Mpixels marketed by Planar Systems, Inc. in its Dome line of products [2].

Spatial dithering   Frame rate control (temporal dithering)

1st 2nd 3rd 4th Frame

Gray level n   Gray level n

Gray level n+1/3   Gray level n+1/4

Gray level n+2/3   Gray level n+1/2

Gray level n+1   Gray level n+3/4

Gray level n+1

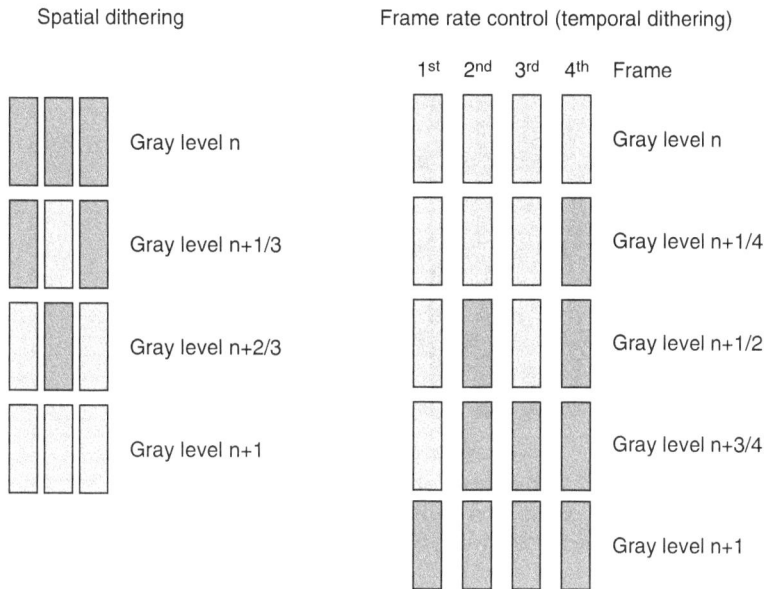

Figure 7.2: Spatial and temporal dithering techniques to increase the number of gray levels in the LCD.

They can be rotated between landscape and portrait mode. The 2-Mpixel monochrome display is also used for referral and clinical review. Ultra-high-resolution LCDs are used in the graphics industry, in satellite mapping, and in military applications as well.

A sophisticated method to increase resolution in color displays without increasing the number of pixels is the use of alternative subpixel architectures and rendering algorithms. An example is the Pentile Matrix™ technology introduced by Clairvoyante [3,4]. The concept is based on a study of how the human eye and brain process visual information. The blue subpixel contributes less than the red and green subpixel to both luminance and the perception of resolution and is therefore de-emphasized in Pentile subpixel designs.

Figure 7.3 shows an example of "logical" pixels with this approach. They consist of a central red or green subpixel surrounded by a blue and four green or red subpixels. The logical pixel has an approximately Gaussian intensity distribution. Each red or green subpixel is used five times, once as the center and four times as the edge of the pixel. It can be shown theoretically and experimentally that the sharing of information between subpixels leads to an effective doubling of the resolution.

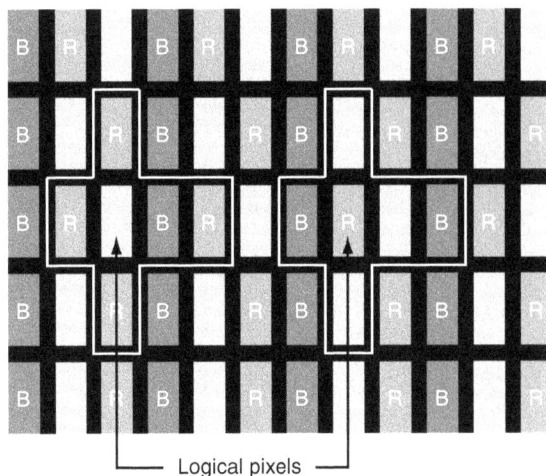

Figure 7.3: Example of Pentile Matrix™ showing logical pixels of a central green or red subpixel surrounded by four other subpixels [3].

Addressing the display requires mapping of the incoming video data to the new subpixel architecture. The increase in resolution was demonstrated with both text patterns and photographic images on prototype displays [3].

The significance of this subpixel rendering method is that with the same design rules, pixel aperture ratio, and number of driver ICs a higher resolution can be obtained. Alternatively, a certain resolution can be maintained with fewer driver ICs and higher pixel aperture and display brightness. First implementation has been on high-end camera phone displays, in which the resolution continues to increase from QCIF to QVGA and VGA formats.

7.2 Reflective and Transflective Displays

In reflective LCDs the backlight is eliminated and power consumption is reduced by more than 60% (Chapter 4, Fig. 4.20). A great deal of effort has gone into the optimization of reflective displays. Since this topic is outside the main scope of this book, it will be only briefly addressed here. For a thorough treatment of reflective LCDs and their performance and issues, the reader is referred to the book *Reflective Liquid Crystal Displays* by S.T. Wu and D.K. Yang [5].

The old, conventional, reflective LCDs in watches and calculators have two polarizers and have a diffuse reflector behind the display. This can lead to parallax in the displayed image, even when the back glass is thin, as illustrated in Fig. 7.4. Reflective LCDs with higher information content use only one polarizer and have diffuse reflectors as pixel electrodes (Fig. 7.4).

An obvious problem with reflective LCDs is that they rely on ambient lighting for legibility. Since ambient lighting conditions can vary widely, both in illuminance and in spectral content, the performance of the reflective LCD will also vary widely. To be readable in low lighting or dark ambient, an auxiliary light source is needed (e.g., a front or side light). Front and side lights with LEDs have been applied in cell phones with monochrome passive LCDs. The performance of pure reflective color LCDs is generally not acceptable, since contrast, luminance, and color gamut degrade much when light has to pass twice through the color filters (coming in and going out). On the other hand, purely transmissive LCDs perform great indoors but their contrast drops off rapidly with increasing ambient lighting, as shown earlier in Chapter 6, Fig. 6.10.

To get acceptable readability under any lighting condition, transflective LCDs have become popular whenever color is needed. They usually have split pixel designs and operate in the reflective mode under high ambient lighting conditions and in transmissive mode in low ambient lighting. An example of a pixel cross section and top view is shown in Fig. 7.5. In the reflective mode, light passes through the color filter and LC cell twice. The LC cell gap in the reflective section of the pixel is therefore about half that of the transmissive section to obtain maximum performance for a certain Δn value of the LC fluid. In addition, the color filter transmission in the reflective section is about half of that in the transmissive section.

When the ambient light level falls below a certain threshold, as measured by a photosensor, the backlight is turned on.

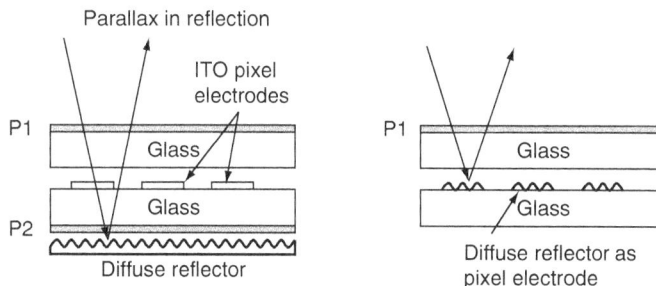

Figure 7.4: Conventional (left) and one polarizer (right) reflective LCDs.

The pixel structure with dual cell gap and dual color filter thicknesses requires more complicated and costly manufacturing processes, but the payoff is an LCD that is legible under all lighting conditions indoors and outdoors. This technology is therefore applied in an increasing number of portable products, such as cell phones, PDAs, digital cameras, and camcorders.

The backlight is only needed in the dark and under low ambient lighting conditions. It can therefore have low luminance and low power consumption.

A well-designed transflective color LCD can maintain acceptable contrast ratio (and color gamut) under any lighting condition up to full sunlight of 100 klux, as shown in Fig. 7.6.

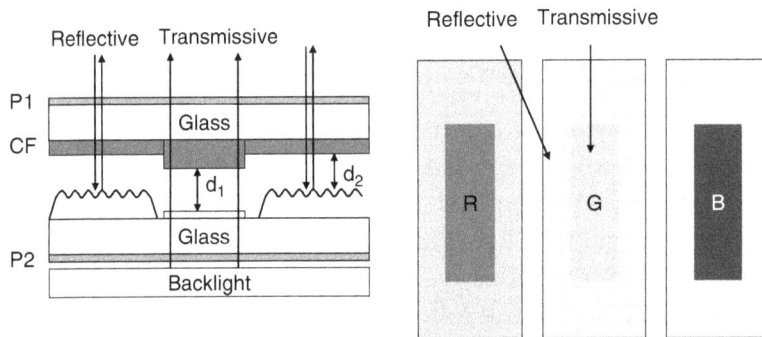

Figure 7.5: Cross, section (left) and top view of color filters (right) of transflective LCD pixel. Cell gaps (d$_1$ and d$_2$) and color filter thicknesses are different for transmissive and reflective pixel areas.

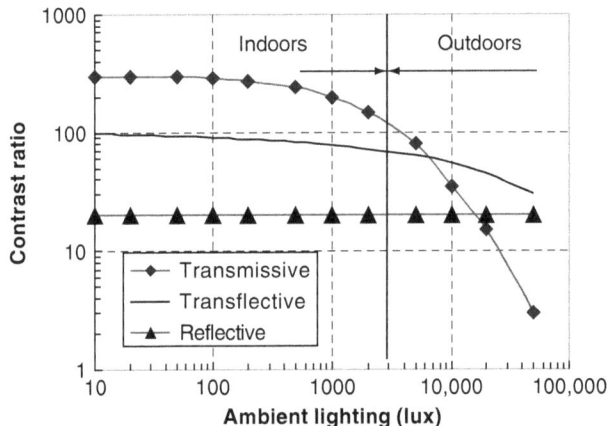

Figure 7.6: Dependence of contrast ratio on ambient lighting for transmissive, reflective, and transflective AMLCDs.

7.3 Field-Sequential Color LCDs

The color filters in a conventional color AMLCD each absorb more than two-thirds of the backlight spectrum. This leads to a significant loss in efficiency and partially explains the poor overall transmission of about 6%.

As explained before, chromaticity in LCDs with RGB color filters is obtained by spatial averaging of the contributions from the three subpixels. There is another method to obtain color, without the use of color filters, which cuts out some of the losses. Instead of juxtaposing the three primary colors in space, they can be sequentially displayed in the same pixel at a fast rate (Fig. 7.7). The backlight flashes red, green, and blue light sequentially, in synchronization with the addressing of the display. This temporal color approach is called a field-sequential color display. The eye will average the three colors over time and perceive the mix as the chromaticity at each pixel.

Field-sequential color displays have been around for a long time, but in LCDs they are quite difficult to implement. The frame is subdivided into three subfields for each primary color, as shown in Fig. 7.8. Each subfield operates at, for example, 180 Hz (three times the frame rate of 60 Hz). The LC needs to have a very fast response time of only a few milliseconds so that the transmittance settles at the final value well within a subfield of 5.5 msec. At the end of each subfield, one of the three colors in the backlight is turned ON.

The merit of this approach is that three subpixels are replaced with one square pixel, opening up the possibility of very high resolution. Along with the elimination of the color filters, this simplifies the top plate and active plate design. In addition, only one-third the number of data driver ICs is required, reducing cost.

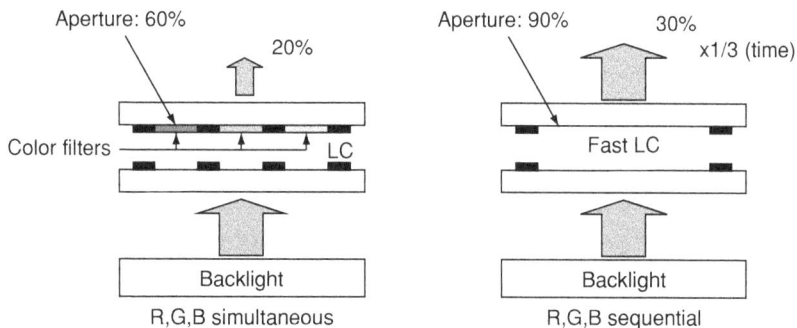

Figure 7.7: Configuration and optical efficiency of conventional (left) and field-sequential (right) color LCD.

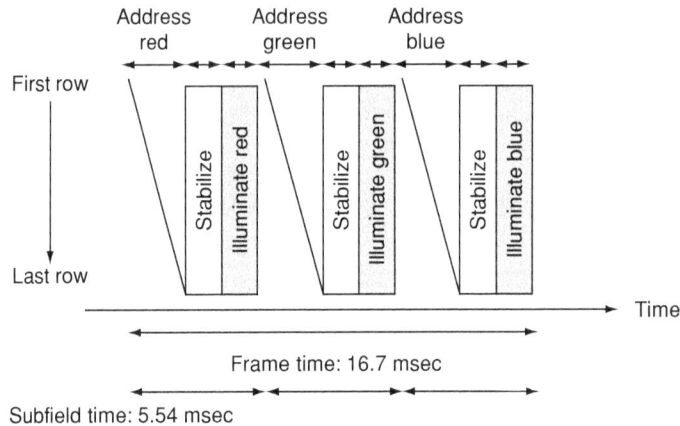

Figure 7.8: Timing diagram for a color field-sequential LCD.

On the down-side, there is the need for a frame memory and image processing to split the video data into separate subfields for each color. A three times higher refresh rate for three times larger pixels implies larger TFTs at each pixel. The backlight is more complicated in the field-sequential mode and has to be synchronized with the vertical scanning of the display. The most difficult problem, however, is to obtain the very fast LC response time of less than 2 or 3 msec. One method to achieve this is to reduce the cell gap to 1.5–2 μm in the TN mode. The response time is proportional to the square of the cell gap (see Eqs. 5.11 and 5.12 in Chapter 5) and will be almost an order of magnitude shorter than in conventional LCDs with 4–5 μm cell gap.

Such a thin cell gap poses serious manufacturing and yield problems, and this is one of the reasons why color field-sequential LCDs have been applied so far only in small displays, such as LCOS microdisplays (see Chapter 8) for near-the-eye viewing and projection and in small handheld LCDs. Other LC modes, such as vertical alignment (VA) with a thin cell gap and ferro-electric and optically compensated bend (OCB) modes with intrinsically shorter response times, are possible choices for field-sequential color displays as well.

The preferred backlight for small handheld units is a combination of red, green, and blue LEDs that can easily and quickly be switched ON and OFF. A very wide color gamut (exceeding 100% NTSC) is possible with the right choice of LEDs. An advantage of LEDs is that they can also be driven at higher current and brightness when the duty ratio is reduced.

The power consumption in field-sequential LCDs has so far not been much different from conventional color LCDs with the same luminance. This can be attributed to the fact that the gain in optical efficiency (see Fig. 7.7) in the display glass is offset by more power consumption in the electronics for higher refresh rate, image processing, and the need for R, G, and B LEDs and their synchronized flashing.

In terms of image quality, color sequential displays can exhibit color breakup of fast-moving images or when the viewer changes his or her focus quickly across the display. The color breakup phenomenon is alleviated by further increasing the refresh rate to 80 or 90 Hz per frame (240 or 270 Hz per subfield).

The market for field-sequential displays is likely to increase in the future with further improvements in design and manufacturing yield.

7.4 Stereoscopic AMLCDs

Three-dimensional display technology provides an opportunity for engineers to enhance the viewing experience by adding the perception of depth. 3D displays can be either based on imaging techniques leading to perspective views or on the presentation of two slightly different images to the left and the right eye. The latter approach is called a stereoscopic display.

Many such systems exist and only a few that are based on AMLCDs, which have recently been introduced, will be described here.

In autostereoscopic displays, the two images are presented to the unaided eye using, for example, a parallax barrier as shown in the example of Fig. 7.9. Odd and even columns of pixels on the display present the two different images to the unaided eye. The viewer does not need special glasses, hence the name autostereoscopic. Since the parallax barrier

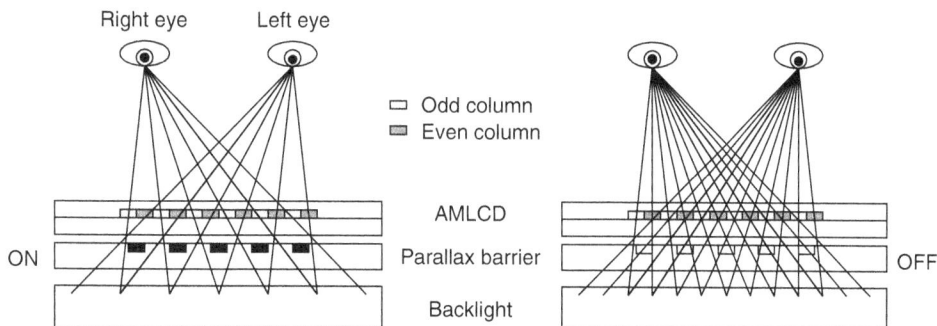

Figure 7.9: Autostereoscopic LCD using switchable parallax barrier for 3D mode (left) or 2D mode (right).

works only over a narrow viewing angle, the viewer needs to position his or her head carefully to experience the 3D visualization.

The parallax barrier can be created with a lenticular screen, with dual directional backlights, or by other methods. In the design of Fig. 7.9, produced by Sharp Corporation, the parallax barrier can be turned ON and OFF to switch between 2D and 3D images with the aid of software [6].

By splitting up the image between adjacent pixels for the left and the right eye, the effective resolution is reduced by 50%, assuming the number of subpixels does not change. This type of display is popular in some higher-end cell phones and has even been applied in notebooks [6]. Extended use tends to lead to eye fatigue, however, and more development is needed to further improve this type of display.

Since the light exiting from most LCDs is inherently polarized, the LCD is a natural candidate for splitting up the stereo video signal into two oppositely polarized images. This can be done temporally or spatially.

In field-sequential stereoscopic displays, the image changes rapidly between the one for the left eye and the right eye. With the aid of polarized shutter glasses, the two images are sequentially presented to the viewer. The electronic shutter glasses are synchronized with the display frame rate and are therefore more expensive than passive glasses. Since these systems require difficult-to-achieve, fast response times for both the LCD and the shutter, they are prone to flicker, even when the frame rate is increased beyond 100 Hz.

Another system for stereoscopic viewing is the stereo monitor with two standard AMLCDs and a half mirror, as shown in Fig. 7.10. This relatively simple design was proposed by Jim Fergason [7], one of the inventors of the TN LCD. The operating principle is shown in Fig. 7.11. The linearly polarized image from one display is presented to the left eye and is transmitted through the half mirror. The image from the other display is reflected by the mirror and presented to the right eye.

The axis of polarization is unaffected in the light path seen in transmission, but is rotated 90 degrees in the reflected light path. The viewer wears polarized glasses with the polarization direction for the left eye and the right eye at 90 degrees. To prevent offsets between the two images, accurate alignment of the mirror and the two LCDs is important. The slightly different images for the left and right eye can be acquired with a dual camera system or can be computer-generated.

This Stereomirror™ system is marketed by Planar Systems, Inc. [8] for 3D medical imaging, 3D aerial mapping, molecular chemistry, and other high-end applications in sizes and resolutions ranging from 17-in. SXGA to 21-in. QSXGA. Gaming is also a

**Figure 7.10: Stereomirror™ display using two AMLCDs and a
half mirror, marketed by Planar Systems, Inc. [2].**

potential market. With the proper configuration of polarizers and mirror, this approach is compatible with the TN LC mode as well as the IPS and MVA modes.

Stereoscopic and 3D displays are considered an area of great interest for future display development and commercialization. A 3D consortium was put together in Japan to promote the research and development of 3D displays [6]. It has worldwide participation with the aim of advancing the technology and use of 3D displays.

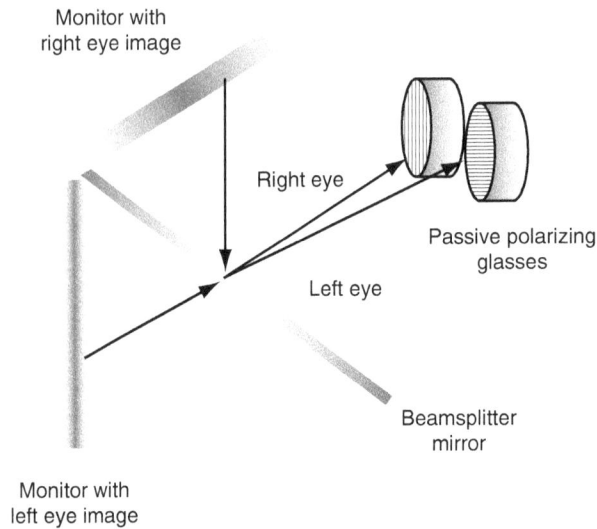

Figure 7.11: Operating principle of the Stereomirror™ display system.

7.5 Touch Screen Technologies

TFT LCDs with touch screens have become popular in displays for industrial and retail applications, as well as in PDAs, smart phones, and tablet PC products. They allow easy and fast access to computer information with a more interactive approach than a mouse or a touch pad. Touch panels use "soft-buttons" that can be easily reconfigured with software. Therefore, they give an integrated solution with more design flexibility than a display with a separate keypad or keyboard.

The majority of touch-enabled AMLCDs are based on resistive, capacitive, inductive, or surface-acoustic-wave touch technology. These solutions require externally added components or screens and are briefly described in the following. In most cases, standard AMLCDs can be used. Many types of touch screens have been adapted from earlier applications in conjunction with CRTs.

The most frequently used and lowest-cost solution is the resistive touch screen [9,10]. It consists of an overlay assembly of two insulating substrates with opposing ITO layers attached to the front of the display. The ITO layers are separated by insulating spacers. Pressure from a finger or some blunt object brings the ITO layers into electrical contact, creating a switch closure.

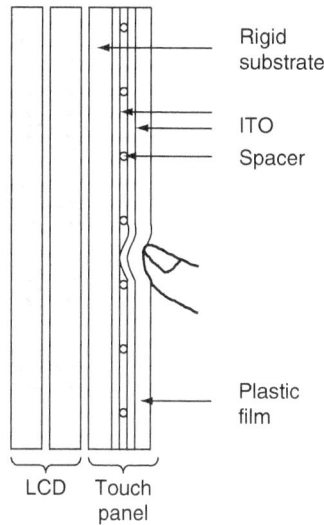

Figure 7.12: Resistive touch panel configuration.

The front insulating substrate is a deformable plastic sheet (Fig. 7.12), while the back substrate is more rigid and can be either glass or plastic.

Resistive touch screens come in several varieties, with four, five, seven, or eight external wires to detect touch location. In Fig. 7.13 the principles of operation of four-wire and five-wire resistive touch screens are shown. In the four-wire approach, a gradient voltage is applied to one surface by the conducting bus-bars at the top and bottom of one ITO layer. When the sheet is touched, the two ITO layers contact each other. The resulting voltage is detected at the other surface and the Y coordinate of touch is calculated with a microcontroller as the ratio of the measured voltage to the total applied voltage. The same sequence is repeated for the other surface by reversing the connections to obtain the X coordinate of touch.

In a five-wire resistive touch panel, a gradient voltage is applied to the bottom ITO surface and detected at the front surface when contact is made between the two ITO layers. One coordinate is calculated again as the ratio of measured voltage to total applied voltage. Then the gradient voltage is applied to the bottom surface with a 90-degree rotation and the other coordinate is calculated with the aid of the microcontroller.

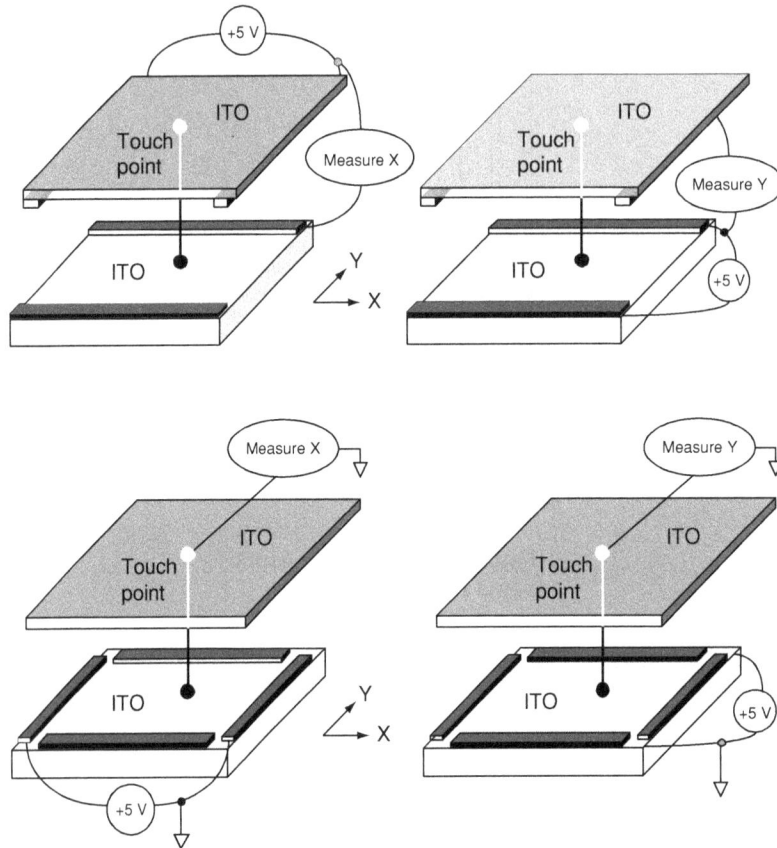

Figure 7.13: Operating principle of four-wire (top) and five-wire (bottom) resistive touch panels.

Resistive touch screens require uniform ITO sheet resistance and initial calibration. Periodic recalibration may also be necessary. They are pressure-sensitive so they respond to any input device, including finger, gloved hand, or pen stylus. Repeated operation can wear out the ITO layers. This technology is dominant in consumer applications where lifetime and durability requirements are not as difficult as in industrial and retail applications.

In Fig. 7.14 the construction of a capacitive touch screen [11] is shown. The sensor consists of an approximately 3-mm-thick bottom glass element and has a conductive sheet on the display side. The ITO on the top side has patterned ITO electrodes on the edges and corners. The touch surface is protected with an overcoat with either a clear or an anti-glare finish.

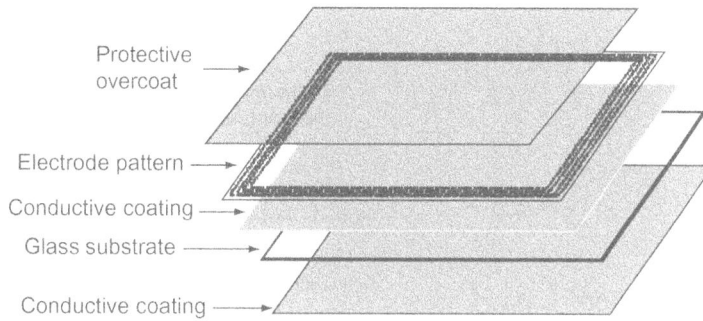

Figure 7.14: Stack-up in a capacitive touch screen.

The electrode pattern uniformly distributes a low voltage field over the conductive layer. When the user touches the surface, there is capacitive coupling to the voltage field. A small current is drawn at the point of contact. The current flow is proportional to the distance from each corner of the patterned electrodes. The controller circuit measures each current flow and calculates the touch location. The capacitive touch screen needs a conductive input device and therefore does not normally work with a gloved hand or plastic stylus.

Application of resistive and capacitive touch screens can be extended beyond simple touch input. With special image processing and software they can be used for signature capture and handwriting recognition. This requires sufficient resolution and fast enough data capture rate of more than 100 Hz. PDAs are examples of this application.

Resistive and capacitive touch screens are add-on overlays on the front of the displays, reducing the display brightness and increasing reflectance as a result of the extra ITO layers and various material interfaces. The touch panels come with several front surfaces: without treatment and with anti-glare, anti-fingerprint, or anti-reflection coatings.

A solution that does not require an overlay screen is the inductive-type touch panel marketed by Wacom Co., Ltd. [12]. Without the add-on panel, the AMLCD maintains its optical characteristics in terms of reflectance and transmittance. The inductive approach uses a sensor board with a grid pattern behind the LCD. The grid switches between sending and receiving mode at a rate of about 50 kHz. The panel is operated with a special stylus with a resonant circuit, consisting of a coil and a capacitor. In the activation mode this resonant circuit is being oscillated (see Fig. 7.15).

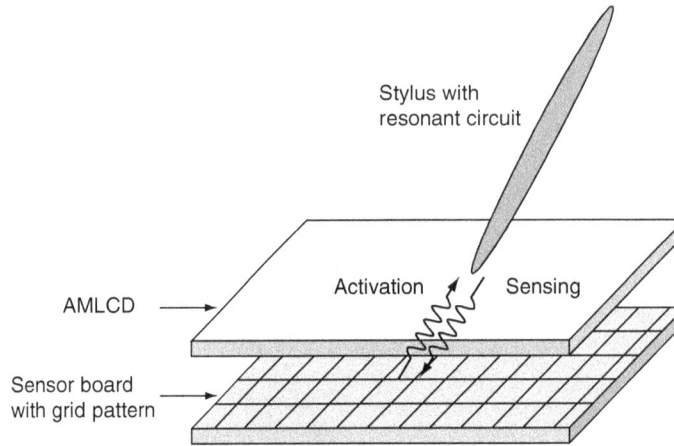

Figure 7.15: Operating principle of an inductive touch panel for tablet PCs [12].

In the sensing mode, the sensor board behind the display analyzes the field being processed through the resonant circuit in the stylus. The signals are analyzed to determine the XY position of the stylus. The passive stylus does not need a battery and can also provide other information such as stylus pressure. In other words, in addition to XY coordinate information, Z-axis forces can also be measured. This technology is dominant in tablet PCs. Handwriting recognition and signature capture are among the possible features. By changing the pressure on the stylus, the width of drawn lines on the tablet PC can be varied as well. The system also senses the proximity of the stylus tip without actual touch, so that moving a cursor while hovering is possible. The capabilities of such stylus-sensor combinations are more advanced than those of simple resistive and capacitive touch panels, which only provide the XY location of physical touch.

One drawback of the inductive approach is the need for a special, relatively costly stylus. Also, the panel cannot be operated with finger touch.

There are other solutions with less image quality degradation as well. An example is the surface-guided acoustic wave (SAW) touch panel [13]. Its construction is shown in Fig. 7.16. The sensor consists of a single 2.8-mm glass element with integrated peripheral wave guides and external piezoelectric transducers. The touch surface is glass with optional AR treatments and overcoat with either clear or anti-glare finish. Conductive surface layers are not allowed.

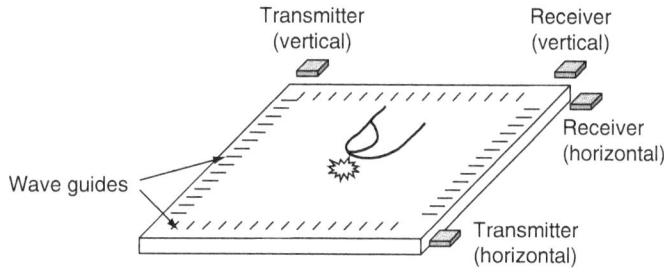

Figure 7.16: Operating principle of a SAW touch screen.

The piezo-electric transducers in the corners are activated by the touch controller. Operation starts with the controller sending a MHz range burst to the transmitter, which converts the signal into ultrasonic mechanical waves across the front surface of the glass. Peripheral guides (reflectors) direct the surface waves. The point of touch contact absorbs wave energy and slows propagation of the wave. This change in the ultrasonic waves registers the position of the touch event and sends this information to the controller for processing. The controller measures time delay and amplitude modulation (i.e., Z axis as well as X/Y coordinates).

SAW touch screens have a high image clarity, since they have an all-glass panel. Without external coatings that can wear out or damage, they are also more durable than resistive screens. The technology is used in public information kiosks, computer-based training, and other high-traffic indoor environments.

3M recently introduced another technology with just a front glass sheet without coatings in front of the display. Labeled "dispersive signal technology" [14], it uses vibration sensors placed at corners behind the glass plate to measure the mechanical vibration created by any object touching the glass. With advanced signal processing, the location of touch can be calculated.

All touch screen approaches have their pros and cons. Resistive technology has the advantage of low cost and is dominant in consumer and portable applications. Its lifetime, however, is limited. Capacitive touch screens are more rugged and have longer lifetimes. They are common in retail and kiosk applications. SAW touch panels are mostly used in high-end applications where cost is less of an issue.

In general, touch screen software control and drivers can be integrated in the operating system similarly to mouse and touch pad control drivers.

References

1. F. Hayashiguchi, "High Resolution Monitors: What Does It Take?" *SID 2001 Digest*, pp. 480–483 (2001).
2. http://www.planar.com
3. C.H. Brown Elliott, T.L. Credelle, S. Han, M.H. Im, M.F. Higgins, and P. Higgins, "Development of the Pentile Matrix™ Color AMLCD Subpixel Architecture and Rendering Algorithms," *Journal of the SID* 11(1), pp. 89–98 (2003).
4. http://www.clairvoyante.com
5. S.T. Wu and D.K. Yang, "Reflective Liquid Crystal Displays," New York: SID Wiley Series in Display Technology (2001).
6. http://www.3dc.gr.jp and www.sharp3d.com
7. J.L. Fergason, "Monitor for Showing High-Resolution and Three-Dimensional Images and Method," U.S. Patent No. 6,703,988.
8. http://www.planar.com/Advantages/Innovation/docs/ds-planar-stereo-mirror.pdf
9. http://www.3mtouch.com
10. http://www.fcai.fujitsu.com
11. http://www.microtouch.com
12. http://www.wacom.com
13. http://www.elotouch.com
14. http://www.dsttouch.com

CHAPTER

8

Alternative Flat Panel Display Technologies

In this chapter, flat panel technologies competing with active matrix LCDs are briefly addressed, including plasma displays, inorganic and organic electroluminescent displays, and front and rear projection. Since the topic of this book is AMLCDs, the discussion of these alternatives will not be more than cursory. These technologies each represent a large field in themselves, with numerous publications and continuing development. The purpose here is to give the reader a flavor of competing technologies and a perspective on the merits and drawbacks of AMLCDs relative to other successful and less successful flat panel display technologies.

The final decision to select a certain display technology is usually based on a careful consideration of price and performance. Since the manufacturing of AMLCDs requires expensive, semiconductor-type processing, it has been difficult to replace CRTs and PDPs with a lower manufacturing cost structure in all applications. Figure 8.1 shows an overview of the major display technologies, of which only the CRT and projection technology are not flat panel displays. LCDs are non-emissive and need external light sources (backlight or ambient light) to be legible. The self-emissive displays can be classified in plasma display panels (PDPs), field-emission displays (FEDs), and inorganic and organic electroluminescent displays.

Field emission displays (FEDs) may be considered a type of flat CRT, in which the electron guns are replaced with very large numbers of microtips in a vacuum cavity. Each pixel contains hundreds of these microtips emitting electrons, which are accelerated by a high electric field to excite phosphors on the opposite glass plate. In the early and mid-1990s, FEDs were touted as a potential replacement for LCDs in notebooks and many other applications. One of the main arguments was the superior viewing angle of FEDs as compared with LCDs. Several companies poured hundreds of millions of dollars into their development. Manufacturing problems turned out to be a major stumbling block. In

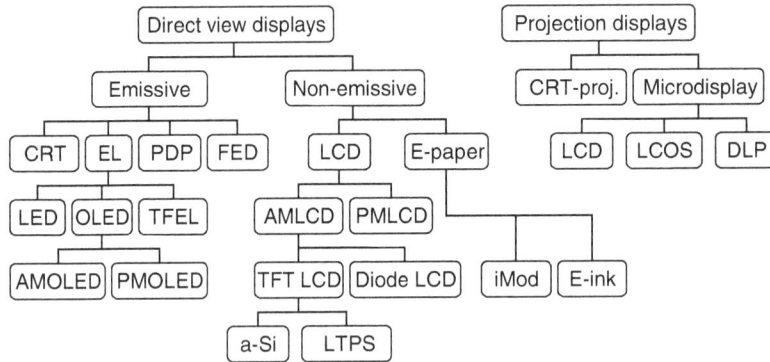

CRT, cathode ray tube; EL, electroluminescent; PDP, plasma display panel; FED, field emission display; LED, light emitting diode; OLED, organic LED; LCOS, liquid crystal on silicon; DLP, digital light projection.

Figure 8.1: Classification of major high information content display technologies.

addition, with continuous viewing angle improvements in AMLCDs, the FED ended up chasing a moving target. This, along with the momentum of the LCD industry, has thus far prevented the commercialization of FEDs. Recently, the research on FEDs has been revived using carbon nanotubes (CNTs) as emitter tips. Since this technology is in the early development stage, it will not be further discussed in this book.

As compared to LCDs, all emissive display technologies, including CRTs, have some distinct advantages and drawbacks. They have generally perfect viewing angle behavior and are self-emissive, not requiring an external light source. They can also have zero luminance in the dark state (by switching OFF the pixel). In a dark room, the contrast ratio of emissive displays can exceed 10,000:1. This compares favorably with LCDs, in which the backlight is permanently ON and contrast ratio is limited, especially at oblique viewing angles. As a result, there will be some residual luminance in the dark state of LCDs.

On the other hand, emissive displays have higher reflectance, rapidly losing contrast at higher ambient lighting levels (see also Fig. 6.10 in Chapter 6). For example, phosphors reflect about 30% of ambient light. To reduce reflectance, contrast enhancement filters such as polarizers or neutral density filters can be added in front of the emissive display at the expense of luminance.

Emissive displays are also more susceptible to burn-in of still images. Both issues, high reflectance and burn-in, are related to the material characteristics (phosphors in CRT, inorganic EL and PDP, organic materials in OLED displays). The luminance of the emitting materials decreases over time. At a given gray level, pixels that have emitted more light after prolonged operation of the panel will have a lower luminance than pixels with a lower emission history. This leads to local image retention (burn-in) of fixed patterns, such as logos and Windows® icons and task bars. With material and design improvements, this problem has been alleviated but not eliminated. The fact that this is not a problem in most LCDs does not mean that the luminance of LCD televisions is constant over their lifetimes of 50,000 hours or more. The backlight in LCDs will deteriorate over time, independent of the ON or OFF state of the pixels. However, this will only cause a global, not a local, decrease in luminance. The LCD uniformity will remain intact after prolonged operation and well-designed LCDs generally do not exhibit burn-in of still images.

8.1 Plasma Displays

Plasma displays have gained acceptance and have captured a large share in the flat panel television market for screen sizes exceeding 40 in. They utilize a plasma of inert gas at each pixel to activate emission from a phosphor. They have a structure as shown in Fig. 8.2, and a pixel cross section as shown in Fig. 8.3.

Barrier ribs on the back glass plate separate channels with red, green, and blue phosphors, deposited on a grid of metal data electrodes. The barrier ribs are usually glass frit patterned by sandblasting. A dielectric layer insulates the data electrodes from the phosphor layers. The front glass plate has transparent scan and sustain electrodes patterned

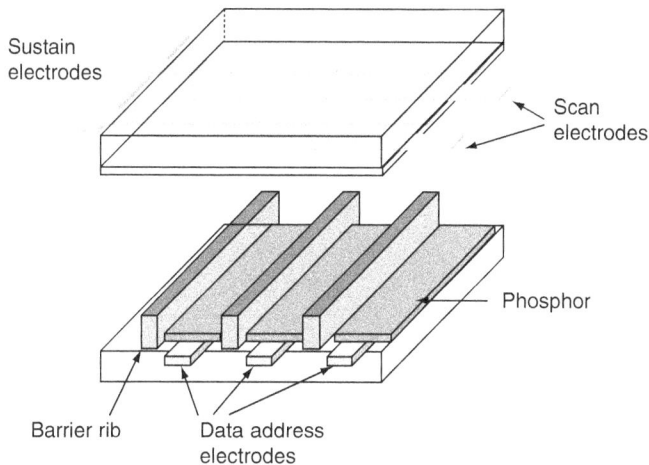

Figure 8.2: Cutout view of a plasma display.

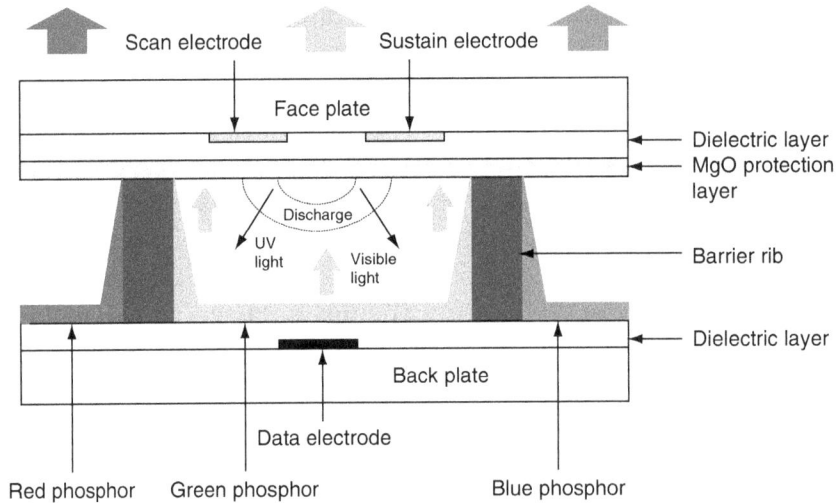

Figure 8.3: Schematic cross section and operation of a PDP pixel.

perpendicular to the data electrodes and separated from the plasma by another dielectric and a MgO protective layer. The sustain electrodes are all connected to the same voltage potential. The panel is hermetically sealed and filled with a gas mixture of Ne with 5–15% Xe at a pressure of 400–500 Torr.

When a row is scanned, the combination of row and data voltage will determine whether the plasma will be ON during a following display period, when sustain pulses are applied to both the scan and sustain electrodes. The Xe gas discharge emits UV light which excites the phosphors.

Most plasma displays are digital (i.e., the plasma is either fully ON or fully OFF). Gray scale is obtained by subdividing an addressing frame of 16.6 msec in subfields, as shown in Fig. 8.4. During each subfield, all rows are sequentially scanned in about 1 msec and the data address voltage will determine if the pixel is ON. Each subfield 1 to 8 has different sustain periods by ratios of 1, 2, 4, 8, 16, 32, 64, 128, respectively. The different combinations of ON and OFF states for all subfields provide all 256 possible gray levels. The incoming digital data signal is modified to address the least significant bit in subfield #1, the second-least significant bit in subfield #2, etc., until the most significant bit in subfield #8.

Both the scan and data driver ICs for PDPs have an output voltage range of about 100 V and are therefore more expensive and have fewer output channels per chip than LCD driver ICs.

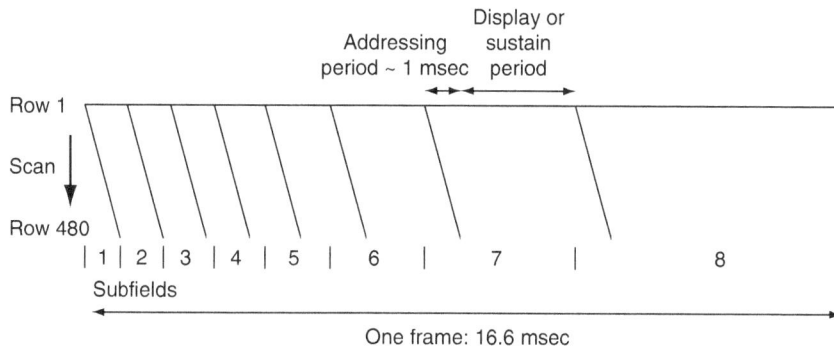

Figure 8.4: Gray scale by pulse width modulation in a plasma display.

Manufacturing of PDP glass panels does not require high-resolution patterning and ultra-clean processing, and is significantly less capital-intensive than AMLCD manufacturing. The size of plasma televisions continues to increase. High-definition plasma televisions with 1080×1920 pixels have been demonstrated with diagonal size exceeding 100 in.

Power consumption in PDPs depends on how many pixels are driven (contrary to LCDs, where power consumption is virtually independent of image content). When all pixels are fully driven, the power is more than twice that of a TFT LCD with comparable size and brightness. The efficiency of PDPs is in the 1 lumen/W range.

8.2 Electroluminescent Displays

Electroluminescent (EL) displays are all-solid-state devices and can therefore be very rugged. They can be subdivided into light-emitting diodes (LEDs) and thin film EL. Inorganic LEDs are based on materials such as GaAs and GaInP. They are well-known from indicator lamps and their efficiency, color range, and luminance have been improved to the point where they are now used in general lighting applications, including automotive lighting and LCD backlights (see Sec. 4.5 in Chapter 4). The excellent brightness and wide temperature range of high-efficiency-power LEDs make them the device of choice in stadium displays, outdoor billboards, and other outdoor applications. Because they are built up from individual LEDs, high information content displays are extremely expensive and limited to niche applications. Unless methods are developed to manufacture monolithic inorganic LED arrays, it is unlikely that they will become mainstream displays.

Organic LEDs (OLEDs), on the other hand, can be processed into monolithic arrays of display pixels and have been successfully commercialized in consumer displays.

The other type of EL device, the inorganic thin film EL (TFEL) display, was one of the early success stories in the flat panel display industry. Because of TFEL's 200-V AC drive requirement and the difficulty in obtaining full-color displays, they were eventually overshadowed by LCDs.

TFEL and OLED displays will be briefly described in the following subsections.

8.2.1 TFEL Displays

Thin film EL direct-view displays are a phosphor-based technology with an inherent sharp threshold, which makes high information content displays possible without an active matrix [1]. They have a structure as shown in Fig. 8.5. Transparent ITO electrodes on glass form one set of addressing electrodes for the data signals. A sandwich of a phosphor layer between two insulators follows. Metal stripes, perpendicular to the bottom ITO electrodes, form the other set of addressing electrodes for selecting the rows. The insulators function to limit the current to the light-emitting phosphor and to prevent shorting. Because of the presence of the insulators, an AC voltage is needed to excite the phosphor.

The phosphor is, for example, ZnS, doped with Manganese (Mn) and with a thickness of 0.5–1.0 μm. At high applied field of about 1.5 MV/cm, the phosphor is excited and starts emitting light from the Mn light emission center. This occurs at about 150 V AC and saturates above 200 V AC. The luminance is approximately proportional to the frequency of the AC driving voltage. Emission from a row of pixels occurs only when the row is selected by the addressing electronics. The passive matrix TFEL display is therefore

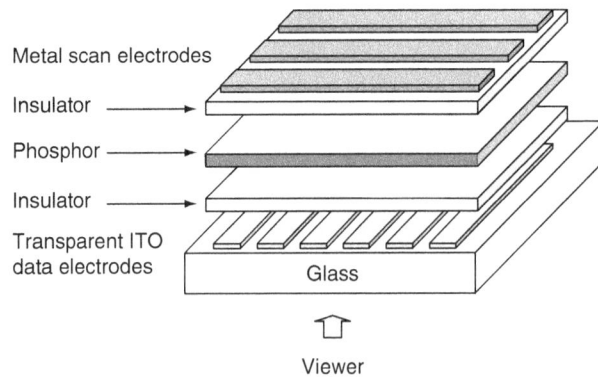

Figure 8.5: Structure of a TFEL panel.

an impulse-type display with good video performance. Row driver ICs with 200-V range and data driver ICs with about 40-V range are needed and a limited number of gray levels can be achieved.

Since the row and column electrodes have to supply the energy for the emission, they need to have low resistance, particularly the Al row electrodes.

TFEL displays have been successfully applied in industrial and medical applications. When using ZnS:Mn, they emit yellow- or amber-colored light. Many other phosphor materials have been studied to obtain emissions with different colors for full-color displays. This has had limited success. The lack of a highly efficient blue-emitting phosphor has been the Achilles heel for this technology and, together with the high voltage drive requirement, has prevented the widespread use of TFEL displays beyond industrial and medical applications.

8.2.2 Organic LED Displays

Efficient organic light emitting diodes (OLEDs) were first reported in 1987 by Tang and Van Slyke of Eastman Kodak [2]. They found that certain organic thin films, when sandwiched between electron injecting and hole injecting layers, emit light with efficiency useful in displays. The physics principle behind this is radiative recombination of electron-hole pairs. Including the electrodes, the film stack consists of five or more thin films optimized to maximize radiative recombination and outcoupling of the light.

After the discovery of organic electroluminescence, a great effort in research and development ensued to improve the luminescence efficiency and lifetime with an eye on display applications.

As an emissive device, the OLED characteristics are quite different from LCDs. They emit light in all directions and they are current-controlled: Their luminance is proportional to the forward current through the diode. The major attractive points of OLED displays are low-voltage operation, the elimination of the backlight, fast response times, and intrinsically very wide viewing angles. This leads to a simpler overall construction with a very thin profile and excellent image quality.

Although the operating principle of OLED displays is quite different, some of the components and thin films used in their manufacture are the same as in LCDs. For example, the same type of glass substrates can be used, and metal and ITO thin films, familiar in LCD manufacturing, form the electrodes for the OLED.

There are two types of OLEDs: small molecule OLEDs and polymer OLEDs, also called polymer light-emitting diodes (PLEDs). Small molecule layers such as Alq3 [tris(8-hydroxyquinoline) Aluminium] are deposited by evaporation in vacuum, while polymer films are spun on a substrate or directly patterned with inkjet printing. Figure 8.6 shows the molecular formula of green-emitting Alq3 along with a typical cross section of an OLED device based on small molecules.

The different OLED and PLED materials can be tailored to emit light at different wavelengths to obtain red, green, and blue colors. The OLED stack has a thickness of only 100–150 nm. Since it covers a large part of the pixel area in a display, care must be taken during manufacturing to prevent shorts between electrodes leading to defective pixels.

In Fig. 8.7 an example of the luminance versus current and luminance versus voltage characteristics is shown for an OLED based on the small molecule Alq3. The luminance increases linearly with current density.

The light emission from both OLEDs and PLEDs degrades over time, more for some colors than others. It leads to differential aging phenomena, with burn-in of static images and color shifts over time. This, along with manufacturing yield, is one of the major issues confronting developers of OLED displays. The decay to half initial brightness depends on material, structural, and drive scheme parameters and is still typically less than 10,000 hours. This has so far limited application of OLEDs mostly to consumer devices that are operated for a small fraction of the time, such as digital cameras, cell phones, MP3 players, and automotive displays.

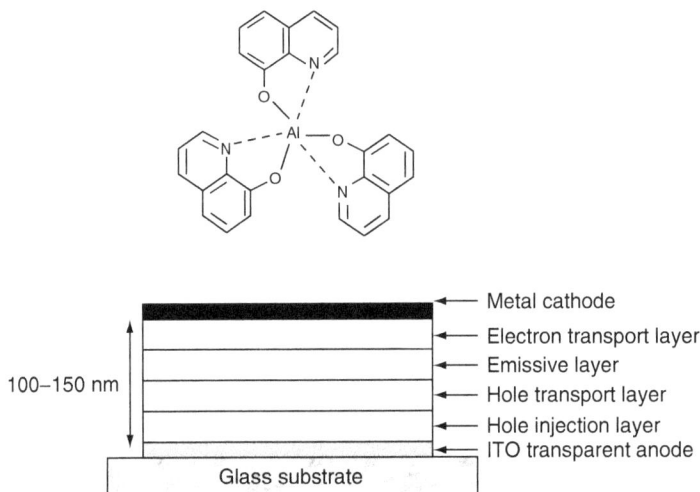

Figure 8.6: Example of a green-emitting small molecule (Alq3, top) and structure of an OLED device (bottom).

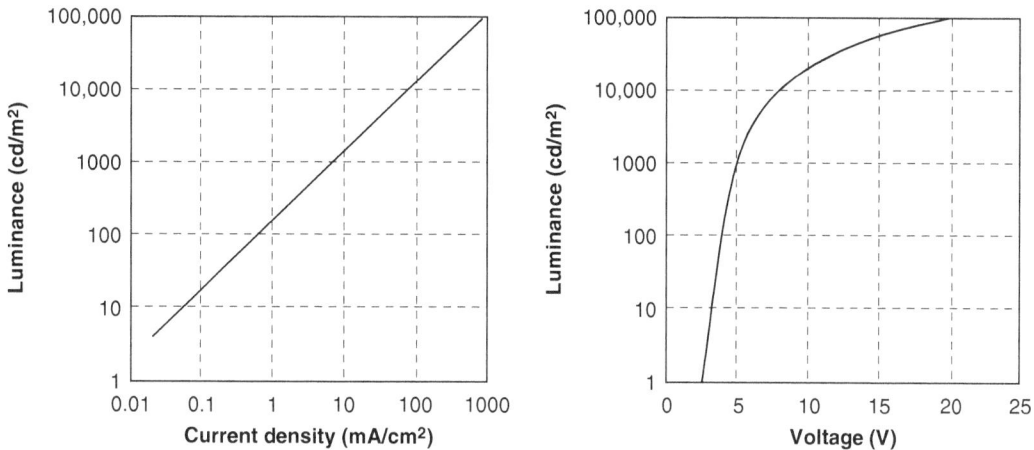

Figure 8.7: Luminance characteristics of Alq3-based OLED.

OLED display development and commercialization are currently among the main focus areas of the display research community. A detailed description of OLED displays is beyond the scope of this book. The reader is referred to the many publications in, for example, proceedings of conferences organized by the Society for Information Display (SID).

8.2.3 Passive Matrix Organic LED Displays

The simplest way to make an OLED dot matrix display is to sandwich the OLED film stack between a grid of transparent column data lines and metal row select lines, as shown in Fig. 8.8. This is the passive matrix configuration. The transparent ITO data lines are deposited and patterned first. They function as the bottom anode electrode for the OLEDs. Since the OLED materials dissolve in the chemicals used for photolithography, the metal top electrodes are patterned without lithography.

Prior to the OLED deposition by vacuum evaporation, a set of parallel bars, called cathode separators, with a re-entrant profile are patterned, typically using a polymer layer. During the subsequent evaporation of the OLED stack and the metal cathode electrodes, the re-entrant profile of the bars separate the metal in electrically isolated stripes, which function as the select lines. Single-color displays are obtained by blanket evaporation of the organic emitter. To achieve color, different organic emitters are evaporated sequentially through a foil shadow mask with narrow slit openings. After the first color deposition, the shadow mask is shifted by the pixel pitch to evaporate the next color emitter.

The display needs to be encapsulated with a metal or glass plate to prevent moisture from degrading the OLED layer. A desiccant is often added inside the encapsulation to ensure acceptable lifetimes.

To prevent specular reflection from the metal back electrodes, a circular polarizer is usually laminated to the front of the display, at the expense of a more than 50% drop in luminance.

Evaporation is not very conducive to high-throughput processing, especially with patterning through a shadow mask. The difficulty of scaling up the process of shadow mask evaporation to large-area substrates has been one of the hurdles for efficient mass production. Therefore, the PLED approach with polymers rather than small molecule organic emitters has received interest. Polymer LEDs are more compatible with large-area processing because the light-emitting polymers can be processed in solvents. For example, they can be deposited by spin coating or by inkjet deposition. The light emission efficiency of PLEDs, however, has so far lagged behind that of small molecule OLEDs.

The passive matrix OLED display is addressed in a one-line-at-a-time method, as shown in Fig. 8.9. The large intrinsic diode capacitance for each pixel is added in the diagram because it plays an important role in passive matrix operation. Only when a row of pixels is selected, the diodes are forward-biased and the OLEDs on that row will emit light. This means that the pixels in a display with 100 rows will emit light not more than 1% of the time. The data driver ICs supply a current signal rather than a voltage signal to control the diode current. Gray scale can be obtained by varying the emission time via pulse width modulation of the data line signal.

Since the duty ratio of passive matrix OLED displays is low, they cannot be scaled up to high-resolution displays with sufficient brightness. Another issue with passive matrix operation is the large intrinsic capacitance (shown in Figs. 8.8 and 8.9) of several pF for each pixel. The large pixel capacitance is a direct result of the thin OLED layer (100–150 nm) and must be charged and discharged every line time. This makes the capacitive load and signal propagation delays unmanageable for any display size larger than a few inches. As a result, passive matrix OLEDs have found application in a limited number of devices, such as in the secondary (outside) display of clamshell cellular phones.

8.2.4   Active Matrix Organic LED Displays

Larger, high information content displays using OLEDs require an active matrix backplane, not unlike the case of LCDs. In an active matrix OLED (AMOLED) display, the

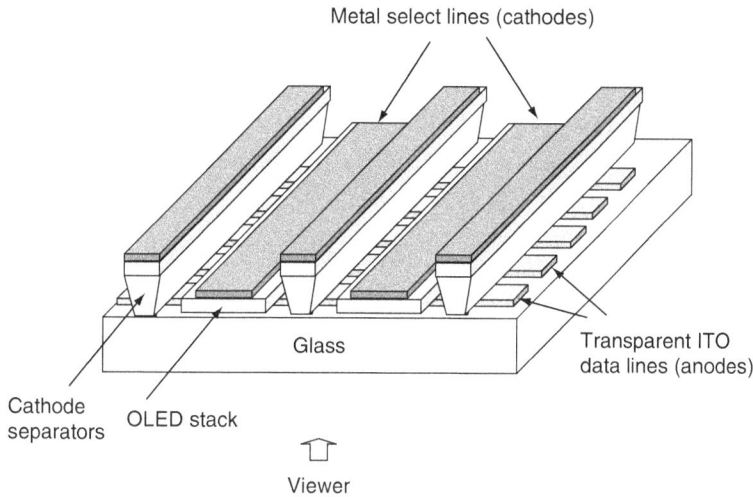

Figure 8.8: Structure of a passive matrix OLED panel.

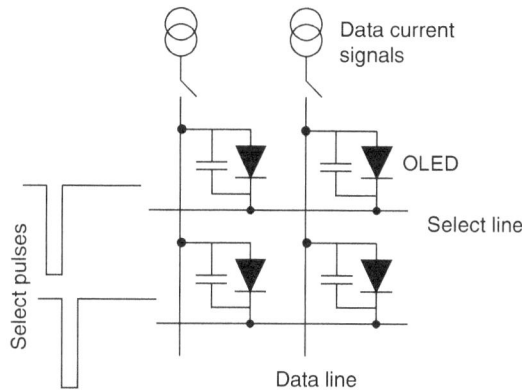

Figure 8.9: Circuit diagram of four pixels in a passive matrix OLED display.

current through the diode at each pixel can be maintained during a large part or most of the frame time, so that the duty cycle of the OLED is much higher. The OLED can therefore be operated closer to its maximum efficiency point. Power consumption is less since the large OLED pixel capacitances are no longer switched at the line frequency.

In contrast to an LC pixel, which is voltage-controlled, the OLED pixel is current-controlled. To address OLEDs with an active matrix backplane requires at least two thin film transistors per pixel. In Fig. 8.10 a circuit diagram of four pixels in the simplest configuration is shown. The access transistors are controlled by the select line and load data voltages to one row at a time. The data voltages are stored on the storage capacitor in each pixel and control the gate of a drive TFT. The drive TFT is connected to a voltage supply and is in series with the OLED. The other terminal of the OLED is the common electrode, a continuous conducting film on top of the OLED, which has the same bias for all pixels (shown as the dotted line in Fig. 8.10).

The data voltage on the gate of the drive TFT sets the current through both the drive TFT and the OLED and, therefore, controls the luminance of the pixel. In the simplest drive scheme, the gate voltage on each drive TFT remains constant during a frame time so that each OLED emits light continuously at a certain gray level during the frame. For moving video images the gate voltage may change from frame to frame.

The drive transistor in the AMOLED display pixel needs to supply a sufficiently large current to the OLED at small source-drain voltage. This ensures that most of the voltage drop and power consumption is in the OLED and not in the drive TFT. In principle, the same type of a-Si TFT and LTPS TFTs used for AMLCDs could be applied in AMOLED displays as well, with some caveats.

Amorphous silicon TFTs have a relatively low mobility and low ON current. A large a-Si TFT is therefore needed to drive the OLED, reducing the pixel area available for the OLED itself. In addition, it was shown in Chapter 2, Sec. 2.5 that the threshold voltage

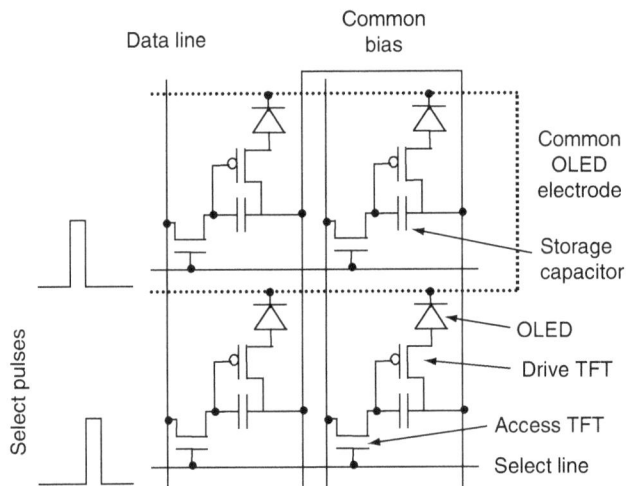

Figure 8.10: Circuit diagram of four pixels in an AMOLED display with two transistors per pixel.

of an a-Si TFT and its ON current are not stable over time, especially when the TFT is kept in the ON state continuously. Unlike pixel TFTs in LCDs, which are used only as ON/OFF switches, the drive TFT in AMOLED displays functions as an analog device and has to accurately control the current through the OLED. The threshold shift in the a-Si TFT will cause a reduction of the OLED current. The a-Si TFT is therefore not very compatible with AMOLED displays unless special pixel circuitry or driving methods are used to compensate for the degradation over time. This is the object of a significant research effort and, if successful, it will take advantage of the large installed base for a-Si TFT array manufacturing.

LTPS TFTs are more stable and can lead to a longer lifetime AMOLED display. They have also a higher mobility that keeps the size and power loss of the drive TFT small and the OLED area in each pixel larger. With LTPS TFTs another problem occurs: the presence of poly-Si grains with different sizes causes variation in the threshold voltages of the drive TFTs from pixel to pixel, which leads to grainy display images. This problem can also be mitigated by different pixel circuits with more TFTs or different drive methods.

In Fig. 8.11, cross sections are shown of bottom-emitting and top-emitting AMOLED display pixels based on an LTPS backplane. In the conventional bottom-emitting configuration, the transparent ITO anode for the OLED is deposited and patterned prior to the OLED layer and the common opaque cathode. The viewer faces the back of the TFT array. This is the easier configuration, since the ITO is a high-quality and highly transparent anode. The pixel aperture is, however, limited by the multiple TFTs and buslines in the pixel. In the top-emitting structure a transparent cathode is used, which is typically a very thin metal layer with a low work function. Its transparency is not as good, but the pixel aperture can be much higher, since the OLED can overlay the TFTs and some of the buslines.

AMOLED displays are an emerging technology and have successfully entered the marketplace in digital cameras and mobile phones. Power consumption depends on how many pixels are lit and, when the highest-efficiency OLED materials are used, it is not much different from backlit LCDs.

The major challenge for AMOLED displays is to further improve lifetimes and manufacturing yields and to reduce cost to compete with established AMLCD technology.

8.3 Electronic Paper and Flexible Displays

In Fig. 7.6 of the previous chapter it was shown that transmissive and reflective LCDs have poor legibility in bright and dim ambient lighting conditions, respectively. One

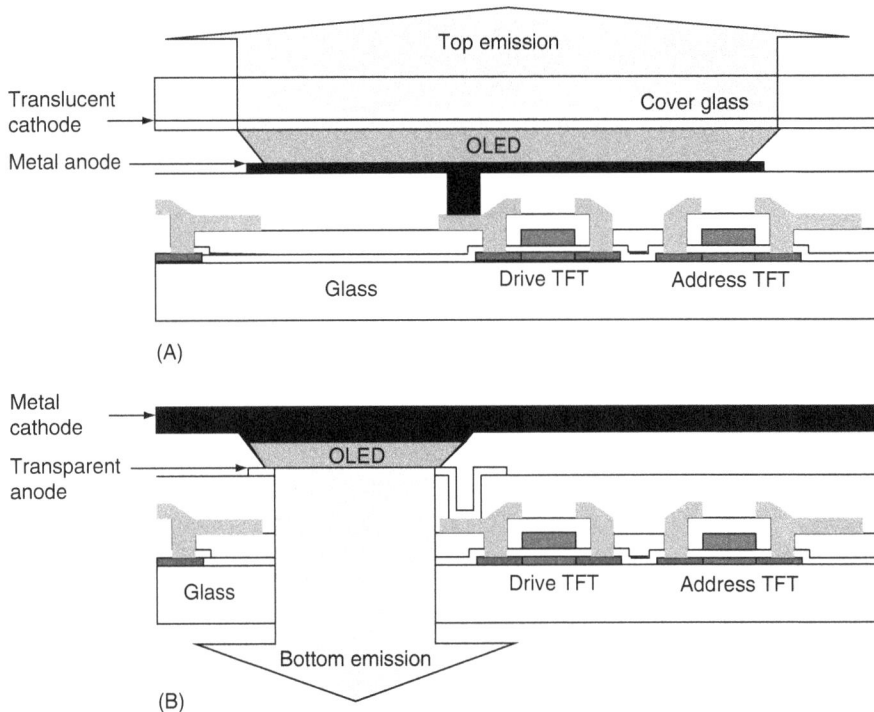

Figure 8.11: Cross section of an AMOLED pixel with top emission (A) and bottom emission (B).

solution for the ambient lighting dependence is the use of transflective LCDs, as outlined in Sec. 7.2 and illustrated in Fig. 7.6 as well.

In high ambient lighting conditions, both reflective and transflective monochrome LCDs reflect less than 15% when the screen is white, because of the limited aperture ratio (fill factor) and losses in the polarizer and other layers. In color versions of transflective LCDs the color filters absorb as well and the reflectance is even lower. This compares unfavorably with the reflection of paper, which is 60–80% (for a white area). Paper retains its excellent legibility over a very wide range of ambient lighting conditions.

When reading documentation or books, most people still prefer paper over an electronic display. This has long been recognized and has spurred the development of alternative display technologies which closer approach the performance of paper. These alternatives can be collectively referred to as "electronic paper." In a review by Drzaic, the issues and state-of-the art of electronic paper have been discussed [3]. A multitude of different LC modes and optical stack-ups have been proposed, some of which are bistable (i.e., they retain the image when the power is switched OFF). Bistable displays need to be addressed only when the information is updated, leading to very low power consumption. Most LCDs, however, have at least one polarizer, which cuts the light by more than 50%. Adding color filters reduces the reflectance even more and makes paper-like appearance impossible.

A number of alternative technologies, not based on LC technology, have therefore been developed. Only a couple of representative efforts will be briefly discussed here: electrophoretic displays and displays based on modulation of optical interference.

Electrophoretic displays have been around for quite some time. E-ink Corporation developed a variation called the microencapsulated electrophoretic display, which has a structure as shown in Fig. 8.12.

Negatively charged white particles and positively charged pigment or dye-containing particles are embedded in a polymer matrix. The microencapsulation in the polymer matrix prevents lateral migration of the particles under gravity and with handling. When a voltage of about +/– 15 V is applied across a pixel, the white particles will move to one side of the cavity and the pigment particles to the other side. The white particles scatter the light and give a paper-like appearance with 30–50% reflectance, while the pigment particles absorb the light to give a dark or color appearance. The viewing angle is excellent and the response time is less than 200 msec. When the power is switched OFF, the last image is retained, making this technology very suitable for applications where a low

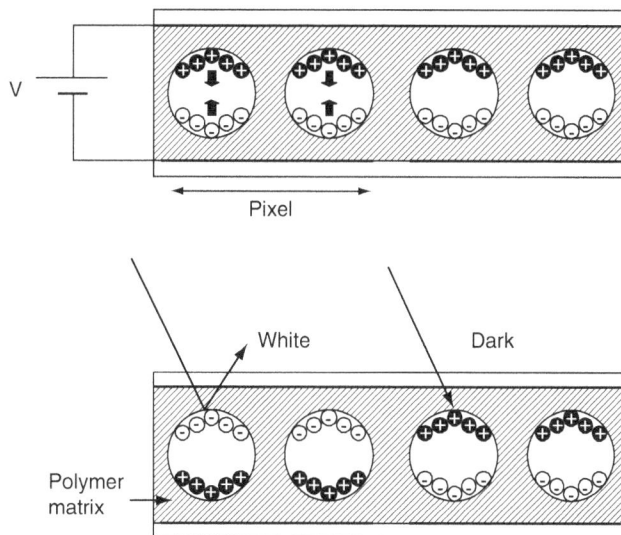

Figure 8.12: Operating principle of the microencapsulated electrophoretic display developed by E-ink.

update rate is acceptable and low power consumption is desirable. Gray scale and color have been demonstrated and the displays can be built on glass as well as on flexible plastic substrates.

For high information content, electrophoretic displays require active matrix addressing. They have successfully been applied in E-books.

A completely different type of reflective display is based on the modulation of optical interference using micro-electro-mechanical systems (MEMS) technology [5]. In Fig. 8.13 the basic structure of the iMod™ is shown, developed by Iridigm Corporation.

Each pixel has a film stack enclosed by a deformable membrane. When the membrane is activated by electrostatic force, it touches the film stack and destructive interference turns the pixel dark. Without the electrostatic force, the stack reflects a color which depends on the air gap between the film stack and the membrane. By properly designing different air gaps, red, green, and blue subpixels are achieved. Usually each subpixel consists of multiple MEMS elements with a pitch of 25–60 μm. The pixels are addressed with low voltage (less than 10 V) and switching speeds are faster than 1 msec. The reflectance versus voltage curve shows hysteresis and, as a result, it is not necessary to use an active matrix backplane to drive the display. Gray scale can be obtained by spatial or temporal modulation and the viewing cone is about 60 degrees, limited by the angular dependence of optical interference. The key feature of this technology is a reflectance of about 30%, several times higher than a color reflective or transflective LCD. The high reflectance, along with low power consumption and the use of standard driver ICs, make this MEMS-type display attractive for portable applications. The technology is being commercialized for small handheld devices, including cell phones.

LCDs, OLED displays, and the technologies described in this section use glass substrates. In some cases, however, there is an incentive to use a flexible substrate, such as plastic or

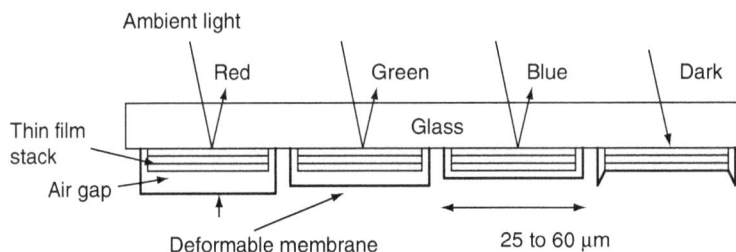

Figure 8.13: Operation principle of the MEMS interferometric display (iMod™) developed by Iridigm Corporation.

stainless steel foil. It reduces the display weight and can make the display more rugged and less likely to be damaged (by breakage). The use of flexible substrates opens up the possibility of roll-to-roll processing for low-cost manufacturing. In addition, a curved form factor or the possibility of rolling up the display is sometimes promoted as an attractive feature for flat panel displays. It should be borne in mind, however, that for CRTs a premium is paid for a flat screen rather than a curved screen.

Much effort has gone into the development of plastic substrates and processing on plastic substrates. Materials such as polyether sulfone (PES) and polyethylene terephtalate (PET) have been used for prototype AMLCDs. Issues with plastic substrates include the need to keep all processing below about 200°C, dimensional instability over time and with temperature, and the difficulty of easily processing bendable substrates. Plastic also tends to be a poor moisture barrier, which requires the deposition of additional barrier films on the substrate. For the latest developments on flexible displays, the reader is referred to the proceedings of several annual display conferences [6].

8.4 Organic Thin Film Transistors

The difficulty of building a-Si or p-Si TFTs on plastic substrates has been one of the main incentives to develop alternative TFTs based on organic materials. Organic TFTs allow low-temperature processing and conform much better to plastic substrates than TFT pixel circuits based on inorganic materials. The latter tend to create stresses on plastic, which pose manufacturing problems and limit the flexibility of the active matrix array substrate. Maximum flexibility and compatibility with plastic substrates is obtained when both the semiconductor and the gate insulator are organic.

In addition to these merits, organic TFTs offer the opportunity to use low-cost, solution-based processes for the various layers, such as inkjet printing and spin coating.

A primary example of an organic semiconductor material is the small molecule material pentacene ($C_{22}H_{14}$). Pentacene consists of five benzene rings with a chain-like aromatic structure. It has a strong tendency to grow ordered molecular crystals. In Fig. 8.14 an example of a cross section for a pentacene TFT is shown, along with the molecular formula of pentacene. The pentacene TFT is a p-type transistor with holes as the majority carriers. The pentacene film can be deposited by thermal evaporation in high vacuum or by spin coating.

As in OLED processing, it is difficult to perform photolithography after the organic material is deposited, since the solvents used in photolithography attack the organic semiconductor film. One way around this problem is to deposit the metal source/drain contacts by evaporation through a shadow mask. With gold source/drain electrodes evaporated on top of the pentacene, hole mobilities as high as 7 cm^2/Vsec have been obtained. However, shadow mask evaporation does not allow small feature sizes, such as a channel length less than 10 μm, needed for TFTs in displays. Therefore, the source and drain contacts in practical applications are usually deposited and patterned prior to the pentacene deposition, as shown in Fig. 8.14.

Figure 8.15 shows the IV characteristics of a pentacene TFT, in which both the polymer gate insulator and the pentacene itself were deposited by spin coating [7]. The field-effect mobility of this particular TFT was 0.02 cm^2/Vsec and was used to build a flexible E-ink display. Row driver circuits using pentacene TFTs have been demonstrated as well. With further improvements in materials, processing, and contact interfaces, other groups have achieved hole mobilities exceeding 2 cm^2/Vsec.

Organic TFTs can be used in AMLCD and AMOLED display arrays, as well as in the backplane for electronic ink displays based on electrophoretics. This is an area of intensive research and development, and for the latest developments the reader is referred to the proceedings of several annual display conferences [6].

8.5   Front and Rear Projection Displays

Strictly speaking, projection displays do not belong to the category of flat panel displays. However, since they compete with PDPs and LCDs in the large television market, they

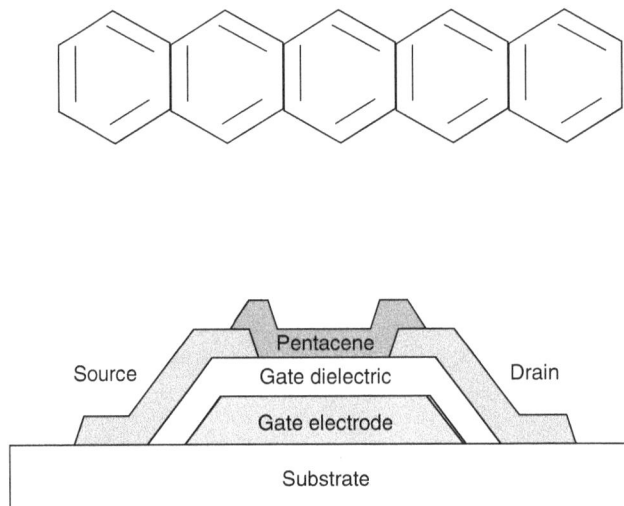

Figure 8.14: Molecular formula of pentacene (top) and a schematic cross section of pentacene TFT.

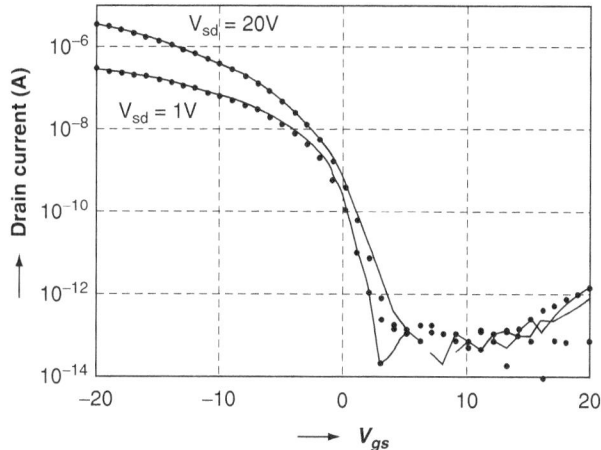

Figure 8.15: Current-voltage characteristics of pentacene TFT with solution—processed gate dielectric and pentacene films [7] (reprinted with permission from the Society for Information Display).

are briefly discussed here. In addition, 60-in. rear projection televisions with a depth of less than 7 in. that come close to the profile of PDPs or LCDs of the same size have been developed. These rear projection displays can be hung on the wall, like PDPs and LCDs.

In projection displays, the electronic display device itself is decoupled from the image screen. Using optics, the image from a small CRT or flat microdisplay is projected on a screen, which is either part of the display system (rear projection) or external (front projection). Both front and rear projectors originally used CRTs. For example, in the passenger areas of large commercial aircraft, systems have been used in which the image of three color projection CRTs are superimposed. The consumer rear projection television market was, until recently, also dominated by systems with three color CRTs. These systems are bulky, heavy, and have limited resolution, brightness, and viewing angle. During the last few years alternative systems based on small light valves have entered the marketplace. The light valves are microdisplays based on LC technology or MEMS. Three types of microdisplays are competing to replace projection CRTs. The first two are based on transmissive and reflective LC technology. The transmissive projection LCD is basically a very high-resolution TFT LCD, most often based on high-temperature poly-Si and sometimes on LTPS. It has part of the electronics integrated on the periphery of the quartz or glass substrate. The LCD can be considered a slide in a slide projector. Since the transmittance of a high-resolution color LCD is not more than 5%, the color filters are normally eliminated in a projection LCD and a color image is obtained with one of a few different methods. To further enhance the transmittance, a microlens array is often added to the light valve assembly to focus the light on each pixel's opening, resulting in a higher effective aperture ratio.

Most LCD-based projection systems use three monochrome, transmissive TFT LCDs, one for each primary color, as shown in Fig. 8.16. The light from the projection lamp is split into three colors by dichroic mirrors and each color beam is modulated by a separate LC light valve. The three color images are then combined and projected by a lens on an external or internal screen. Brightness from a projector is expressed in ANSI lumens, the total integrated light intensity. Typical values are 500–8000 lumens.

The second approach uses reflective LCDs on CMOS chips, called liquid crystal on silicon (LCOS). The great advantage of LCOS is that the front-end production (i.e., the CMOS circuitry) can be done in standard IC factories with very well-established wafer processes. After fabrication of the chips with pixel array and peripheral electronics, the semiconductor wafer is subsequently covered with an ITO-coated glass plate, filled with LC in TN, VA, or other mode, and cut into individual devices with a diagonal size of about 1 in. Polarizers are laminated to the individual microdisplays and a flex connector is bonded to supply input signals and power. An example of an LCOS pixel cross section is shown in Fig. 8.17. The pixel circuit is built up from CMOS circuitry and includes a

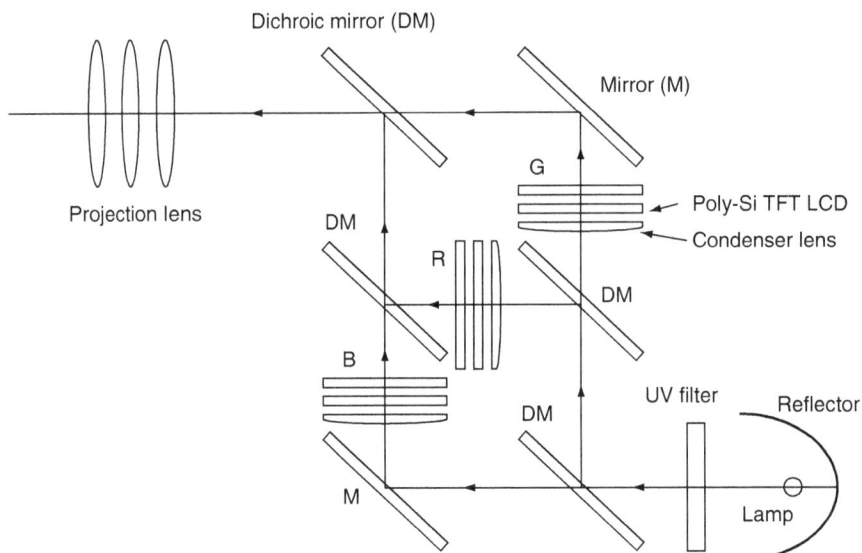

Figure 8.16: Operating principle of a projection system with three transmissive poly-Si TFT light valves.

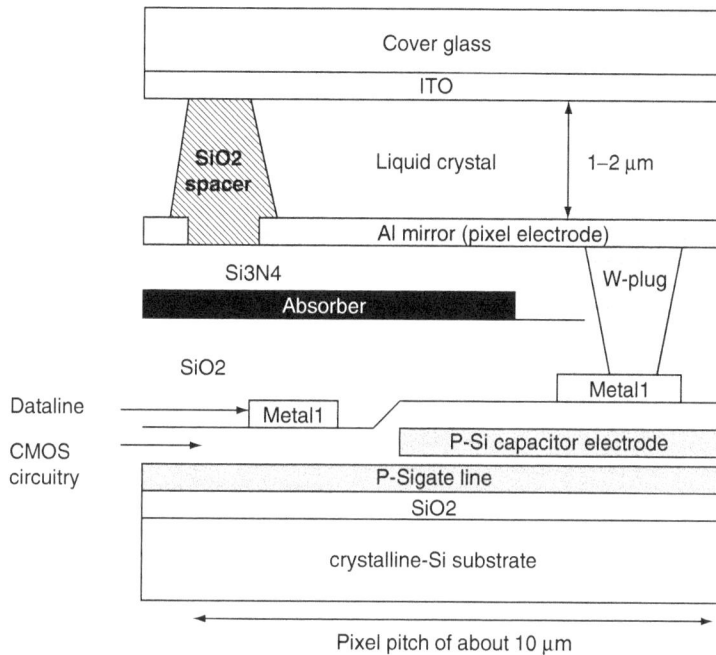

Figure 8.17: Example of a pixel cross section in an LCOS projection microdisplay.

relatively large storage capacitor. Photocurrent in the transistors is prevented by the patterning of an absorber layer.

The pixel electrodes are made from highly reflective Al, overlaying most of the pixel circuitry. A thin front glass cover plate with a blanket ITO coating for the common electrode is bonded to the chip with a gap of typically 1–2 μm, controlled by SiO_2 spacers.

Color in LCOS projection systems can be achieved with three such reflective light valves. Alternatively, the LCOS chip can be operated in a color field-sequential mode. When the cell gap is 1–2 μm, the LC response time is fast enough to switch the image in a few msec. It is then possible to use a single LCOS device in combination with a color wheel for a color display by subdividing the frame time into three subfields for the primary colors. This single-light valve system leads to lower-cost projection television, but also to lower brightness. The optics system is significantly simpler because accurate alignment of three light valves for superposition of three images is no longer needed. Polarization recovery methods have been proposed to improve brightness by recycling the lost polarization component. LCOS is also used with different LC modes, which have an inherently much faster response time (for example, the ferro-electric LC mode). Figure 8.18 shows an example of the projection optics for a system with three LCOS chips.

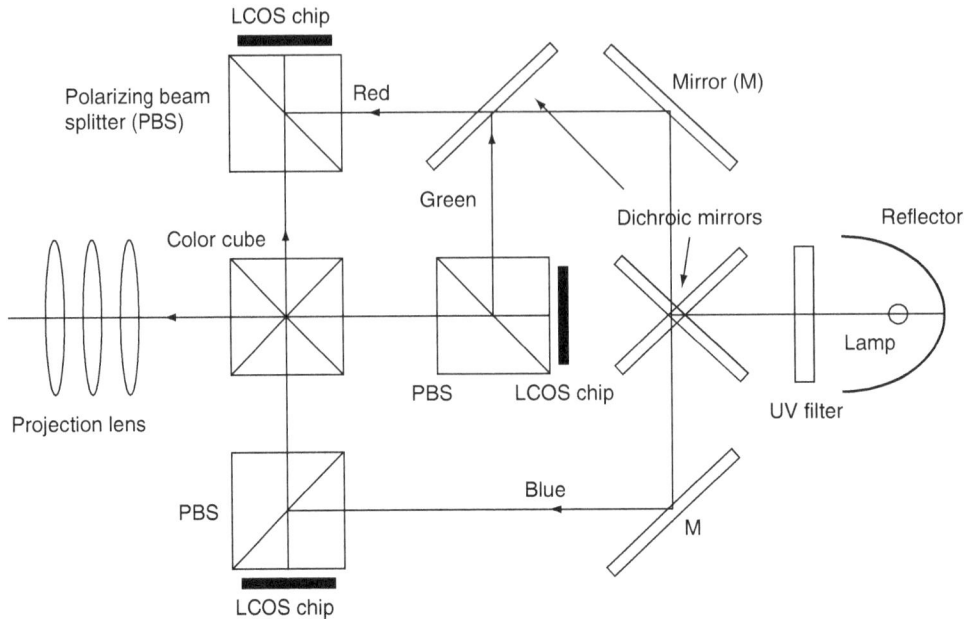

Figure 8.18: Projection system with three reflective LCOS chips.

The third technology is a MEMS system using a digital micromirror device (DMD™) developed by Texas Instruments [8]. It consists of an array of micromirrors on a CMOS chip and forms the basis of TI's digital light processing (DLP™) technology (Fig. 8.19). The mirrors can be tilted by +10 degrees or −10 degrees by electrostatic potentials applied to the mirror structure at a maximum rate of more than 1000 times per second. In one of the tilted positions the light is directed toward the optical system and projected on the screen. In the other tilted position, light is deflected out of the system. By switching at a rapid rate, a gray scale of up to 10 bits can be achieved at 60-Hz operation in this 100% digital system. As in projection LCDs, a color image can be obtained from combining the images of three DMD™s, one for each primary color. Another option is to use a single DMD™ with a color wheel. The color wheel rotates so that the three primary colors illuminate the DMD™ sequentially. Red, green, and blue images are superimposed in rapid succession on the screen from a single device, resulting in a field-sequential color image.

The DMD™ has the advantage that is does not rely on polarized light, which more than doubles the light emanating from the projection device. On the other hand, the optical engine for deflection of part of the light is somewhat more complicated than the projection optics for LCD-based projectors.

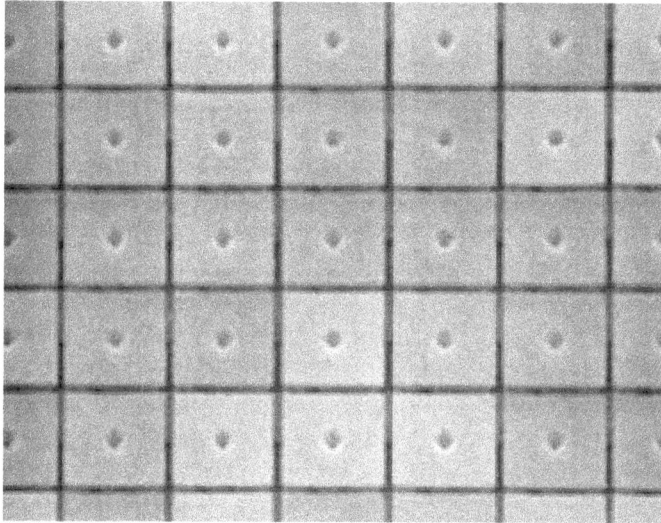

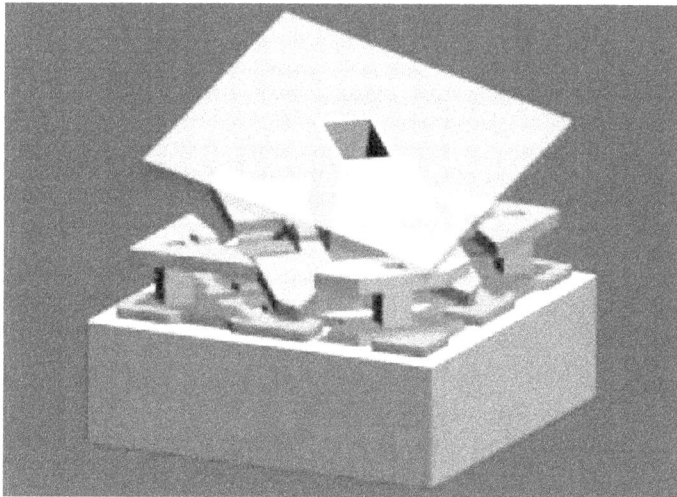

Figure 8.19: Photograph of pixels (top) and structure of one pixel (bottom) in a Texas Instruments DMD™ device [8].

Systems with each of these three microdisplay technologies have been commercialized for front and rear projection applications, including high-definition television. The DLP™ technology, in particular, has captured a large share of both the rear and front projection display markets. By folding the projection optics in a thin profile, rear projection televisions with less than 7-in. depth are now on the market [9]. Figure 8.20 shows the generic configuration of a rear projection television based on LCD panels.

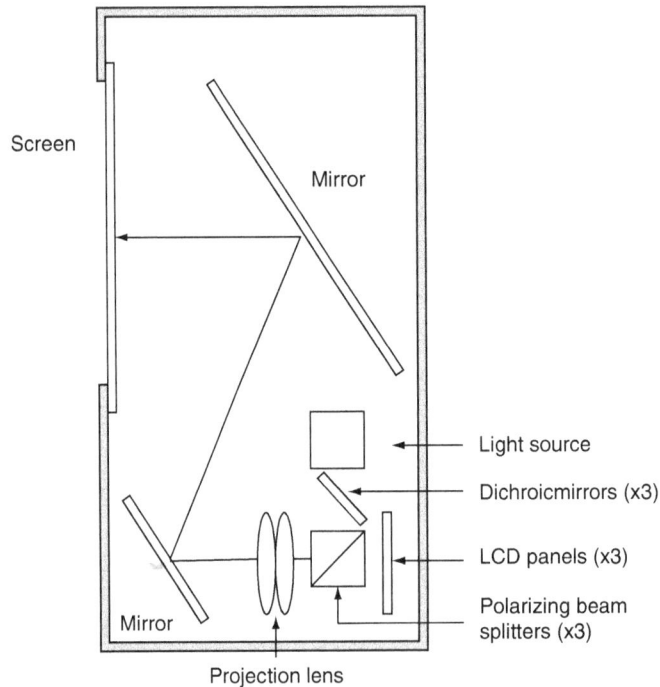

Figure 8.20. Example of a rear projection system based on LCD panels.

An important component in projection systems is the light source. Xenon, metal halide, and ultra-high-pressure tungsten arc lamps are used. The lamps are being improved in brightness and have progressively longer lifetimes to meet the specifications of front and rear projection televisions. Ideally, lamp replacement would not be needed over the lifetime of the system.

The impressive efficiency improvements in high-power LED lamps have captured the attention of projection system designers as well. A very compact RGB LED-based illuminator for pocket-sized projectors was introduced by Lumileds Lighting [10]. It is designed for a single light valve operating in the field-sequential color mode. As a personal projection device, it could be a companion product to a digital camera or a PDA.

It should be noted that some of the microdisplay technologies described here are also used for near-the-eye applications, such as head-mounted displays. These systems employ one or two microdisplays, for one or both eyes. When used in pairs, they provide a possibility for stereoscopic viewing.

References

1. C.N. King, http://www.planar.com/Advantages/WhitePapers/docs/ELD_200307.pdf
2. C.W. Tang and S.A. van Slyke, "Organic Electroluminescent Diodes," *Appl. Phys. Lett.* 51, pp. 913–915 (1987).
3. P. Drzaic, "Electronic Paper: The Quest for the Killer Application," Application Tutorial A-1, *SID 04 Application Tutorial Notes*, pp. A-1/1–45 (2004).
4. http://www.eink.com
5. http://www.iridigm.com
6. http://www.sid.org
7. P.J.G. van Lieshout, H.E.A. Huitema, E. van Veenendaal, L.R.R. Schrijnemakers, G.H. Gelinck, F.J. Touwslager, and E. Cantatore, "System-on-Plastic with Organic Electronics: A Flexible QVGA Display and Integrated Drivers," *SID 2004 Digest*, pp. 1290–1293 (2004).
8. http://www.dlp.com
9. http://www.infocus.com
10. M.H. Keuper, G. Harbers and S. Paolini, "RGB LED Illuminator for Pocket-Sized Projectors," *SID 2004 Digest*, pp. 943–946 (2004).

Active Matrix Flat Panel Image Sensors

In this final chapter, flat panel technologies using similar types of active matrix arrays, as in AMLCDs, are described. They underscore the more general application of a-Si TFT arrays in electronic devices. A prominent one is their use in 2D image sensors for X-ray detectors. These detectors are replacing X-ray film and, combined with ultra-high-resolution AMLCDs, provide diagnostic-quality images to radiologists. They eliminate all chemical film processing and allow easy electronic storage and transfer of digital image files.

9.1 Flat Panel Image Sensors

Instead of writing data to an active matrix array, as is done in displays, it is also possible to read data from specialized active matrix arrays that function as large-area image sensors. They have a photo-conversion material and/or photosensitive device added at each pixel. The resulting flat panel sensor is scanned one row at a time, in a similar way as AMLCDs. The data driver ICs are replaced by readout chips which amplify the readout charge from each pixel and multiplex the data signals from the array into a video signal, representing the 2D image. This technology is now increasingly used for flat panel X-ray detectors in medical and industrial imaging and in security applications.

Discovered by Röntgen in 1895, projection radiography is the oldest medical imaging technique; more than 200,000 systems are in use worldwide. Since there are no good optics elements available to focus an X-ray image, the X-ray system uses projection from the X-ray source through the object onto a full-field-size receptor, as shown in Fig. 9.1. Historically, the receptor has been a screen-film combination. The screen-film combination consists of a light-sensitive silver-halide film sandwiched between two radioluminescent screens or scintillators. The scintillator can be, for example, Gd_2O_2S:Tb powder in a binding agent. It absorbs X-ray photons and converts them into visible light, which exposes the silver-halide film. The light-sensitive film needs to be developed with chemicals and, after review, it is stored in large hospital storage rooms. Radiologists usually examine X-ray films by placing them on high-brightness light boxes.

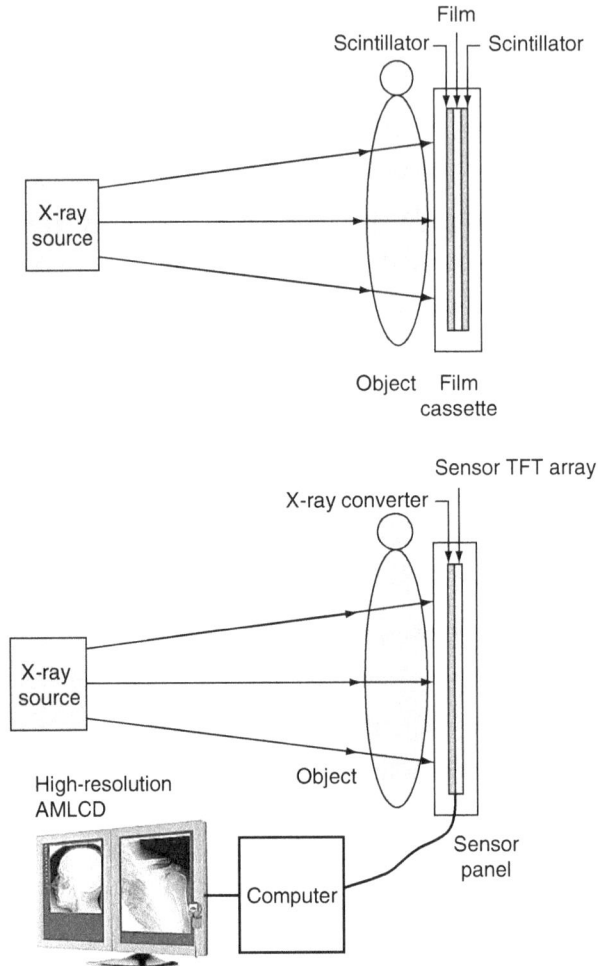

Figure 9.1: Projection radiography with screen-film combination (top) and with flat panel detector (bottom).

The flat panel active matrix detector replaces this film-screen combination and high-resolution, monochrome AMLCDs, such as described in Chapter 7, Sec. 7.1, replace the X-ray film on a light box. The bottom half of Fig. 9.1 schematically shows the X-ray detector system, including two monochrome LCDs. Such a system is often referred to as a filmless, direct radiography system.

There are two types of flat panel X-ray detector technologies:

1. Direct detectors, which convert X-rays absorbed in a photoconductor directly into charge. This is called the photoconductor or direct approach.

2. Indirect detectors, which first convert X-rays to visible light in a phosphor screen or scintillator and then detect the visible light in a photosensor array. This is called the scintillator or indirect approach.

The indirect technology with a scintillator screen is also used in conjunction with a high-resolution CCD or CMOS image sensor. In this case, the visible light generated in the scintillator is first focused by optics on the much smaller image sensor chip. This third approach has been commercially successful as well. However, there are losses of resolution and of light efficiency in the focusing optics and the overall system is more bulky, which has prevented its widespread use.

Flat panel detectors have the same thin profile as a screen-film cassette and can therefore, in principle, be a drop-in replacement for the cassette. Since the TFT at each pixel in a flat panel detector functions as a simple ON/OFF switch at low switching speed, the a-Si TFT has been perfectly adequate for this application.

9.2 Direct Conversion Detectors

In Figs. 9.2 and 9.3, a schematic cross section and a circuit diagram of a direct conversion X-ray sensor are shown. The sensor array consists of a TFT matrix with a storage capacitor at each pixel and row and column lines, comparable to a TFT matrix for LCDs. The array is coated with a thick layer of X-ray photoconductor, for example, 500–1000 μm of Selenium (Se), which absorbs most of the incident X-rays [1,2].

On top of the Se, a metal electrode layer such as a thin Al film is coated, which transmits the X-rays. During an image acquisition, a high positive voltage of up to 6 kV is

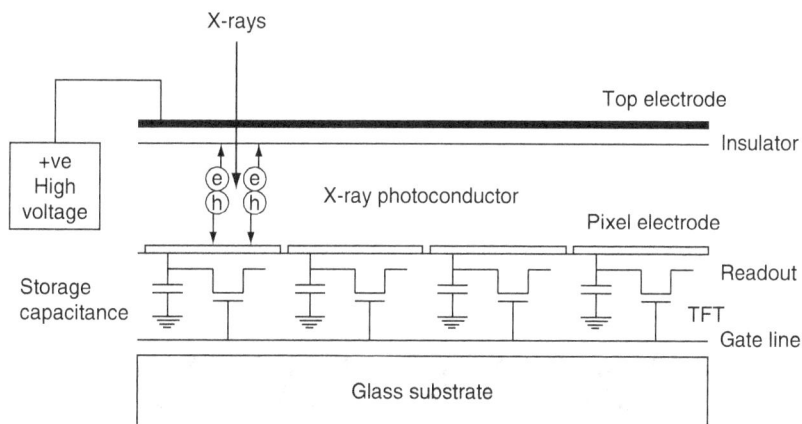

Figure 9.2: Schematic cross section of direct conversion X-ray detector.

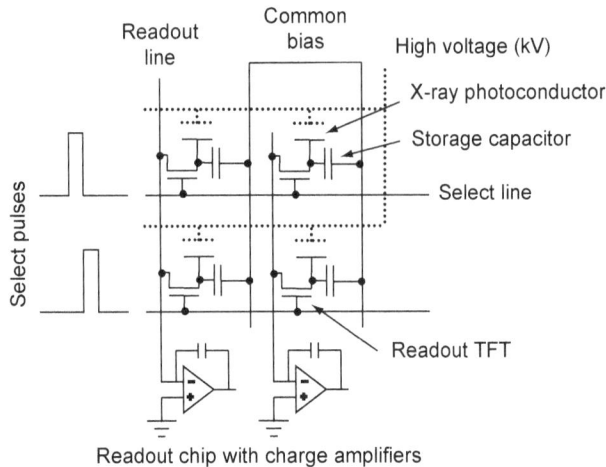

Figure 9.3: Circuit diagram of four pixels in a direct conversion X-ray sensor.

applied to this top electrode. Each X-ray photon, when absorbed in the photoconductor, creates more than 1000 electron-hole pairs. The high electric field across the Se layer separates the electrons and holes. The holes are collected on the pixel electrodes so that positive charge proportional to the number of absorbed X-ray photons accumulates on the storage capacitors (also called collection capacitors) of the TFT array. After each exposure the charge is read out one row at a time by sequentially turning ON the readout a-Si TFTs of each row, using externally connected row select driver ICs. These driver ICs are similar to the ones used in AMLCDs.

X-ray exposure times are typically a few 100 msec and the panel is read out in about 1 second.

External readout chips with charge amplifiers at their input are connected to the column pads by wire bonding or TAB bonding. The output signals are multiplexed into a serial video signal and then digitized into 12- or 14-bit digital images by A/D converters.

The image is subsequently processed with offset and gain correction to cancel out the dark current noise and fixed pattern noise. Processed images can be displayed on a monitor (such as the medical imaging AMLCDs described in Chapter 7, Sec. 7.1 and shown in Fig. 7.1). They can also be stored on computer media and transferred to other locations on an intranet or the Internet for examination, review, and remote diagnostics. Direct conversion X-ray detectors have been successfully applied in general radiography and mammography. The detectors for general radiography have a size of 14 in. × 17 in., similar to the size of standard X-ray film, and square pixels with a pitch of 127–150 μm. This portrait size is sufficient for chest X-ray images. To avoid the need for rotation of the X-ray detector for imaging of some other body sections such as extremities, several manufacturers also market 17-in. × 17-in. sensor panels. For mammography, smaller panels of 7 in. × 10 in. are typical.

Figure 9.4 shows a pixel layout and pixel cross section for a typical direct radiography sensor. The collection electrode (ITO #2) must cover a large part of the total pixel area (> 90%) to achieve maximum charge collection. The collection electrode therefore overlaps the buslines in a similar fashion, as in super-high-aperture AMLCDs described in Chapter 6, Sec. 6.1.2. As in LCDs, the interlevel dielectric is preferably a thick polymer to minimize the data line capacitance, which contributes to kTC noise. In fact, the

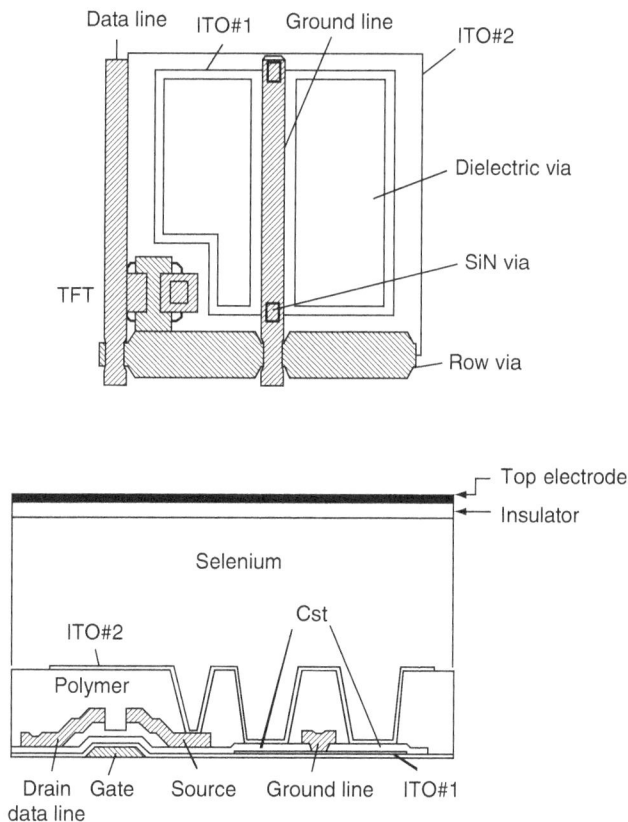

Figure 9.4: Pixel layout and schematic pixel cross section of a direct X-ray sensor pixel showing readout TFT and storage capacitor C_{st}.

process for the a-Si TFT array in direct X-ray sensors can be very similar to that for high-aperture TFT arrays [3] and they can therefore be produced in the same factories.

The TFT array needs to be protected from the very high voltage (in the kV range) applied to the top of the Se layer. One way of doing this is to add an insulator layer between the top electrode and the Se X-ray photoconductor. This avoids damage to the TFTs when there are pinholes or weak spots in the Se or when the voltage across the photoconductor collapses at high X-ray exposure. The thickness of the insulator layer is optimized so that its capacitance is small and the voltage across the storage capacitor of each pixel cannot exceed 20 or 25 V, even at maximum X-ray exposure, when the voltage across the photoconductor collapses.

A drawback of adding the insulator on top of the Se is that charge trapped in the Se during X-ray exposure cannot freely dissipate. To reset the Se after each exposure a backlight flash is therefore used, which detraps the charges. This "optical reset" is only possible when the panel is mostly transparent, hence the transparent ITO electrodes for the storage capacitor.

Another important aspect is the dark current of the Se layer (in the absence of X-rays). The interface between the ITO collection electrode and the photoconductor should block the injection of electrons into the photoconductor. An interface blocking layer is added for this purpose.

Amorphous Se photoconductors have successfully been applied in flat panel sensor products, as an extension of their use in xerography. However, they have some drawbacks, including very high voltage requirement (up to 6 kV) and very large thickness (up to 1000 μm, as thick as or thicker than the glass substrate). Another issue with Se is that it tends to starts crystallizing at temperatures higher than about 50°C, which deteriorates performance. This makes handling and transportation a major issue. Figure 9.5 shows an X-ray image of a foot, acquired by a direct conversion flat panel detector with Se photoconductor. Details of the bone structure are clearly visible.

To address the drawbacks of Se, other X-ray converters with similar or higher X-ray stopping power and lower operating voltages are under investigation, including PbI_2, CdTe, HgI_2, and PbO.

9.3 Indirect Conversion Detectors

In Figs. 9.6 and 9.7, the schematic pixel cross section and circuit diagram of an indirect X-ray sensor are shown. The active matrix array has a readout a-Si TFT and an amorphous silicon PIN photodiode at each pixel. Since the TFT and the photodiode are

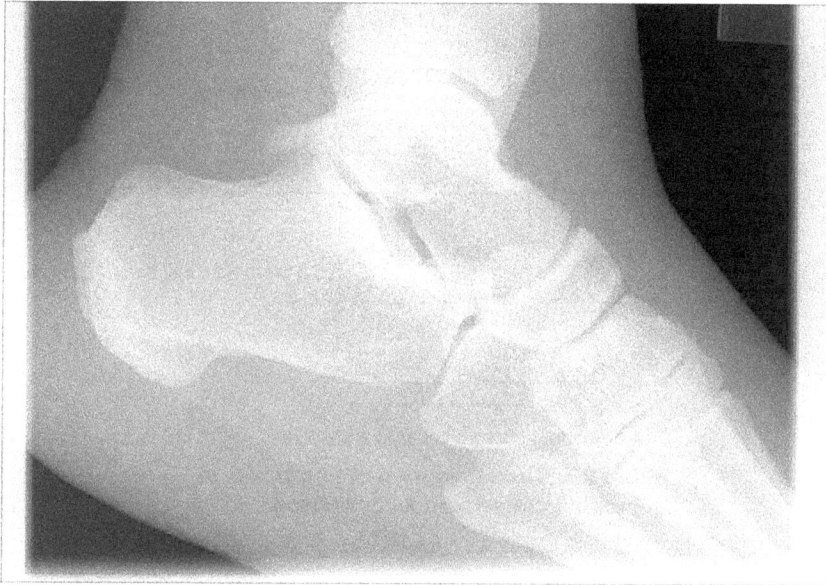

Figure 9.5: X-ray image of part of a foot, acquired by a direct conversion flat panel detector.

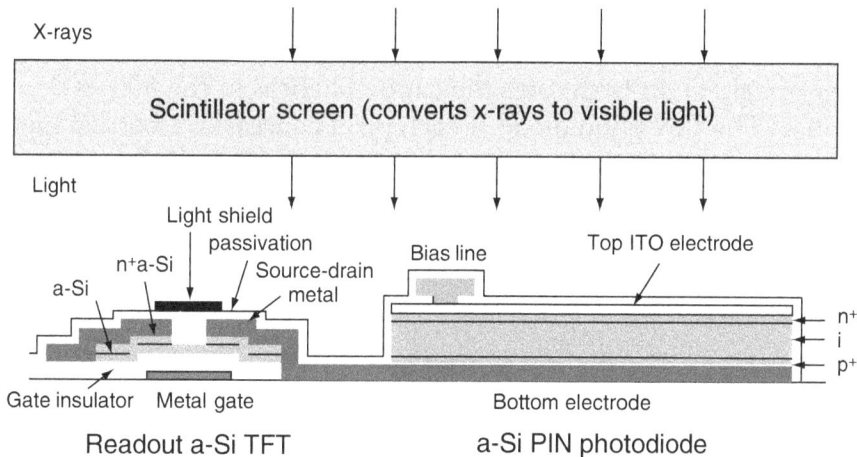

Figure 9.6: Schematic cross section of a pixel in an indirect conversion X-ray detector.

made in different process steps, the total number of mask steps for this type of active matrix detector can be as high as 13. The extra processing for the PIN diode makes it difficult or impossible to process the array on standard TFT lines used in the AMLCD industry.

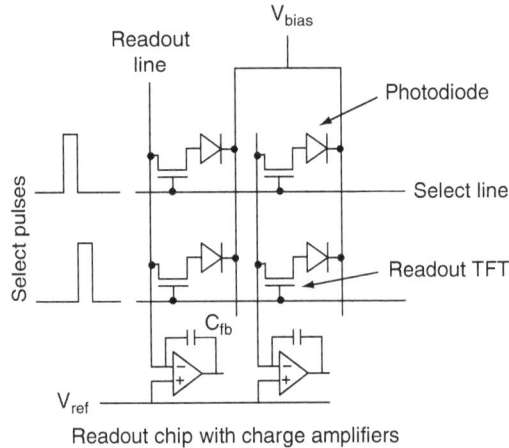

Figure 9.7: Circuit diagram of four pixels in an indirect conversion X-ray sensor.

In this case, the X-rays are first converted in an overlying scintillator or radioluminescent screen into a visible glow. The visible light is absorbed in the photodiode. The scintillator can be a Gd_2O_2S:Tb screen, such as is used in screen-film combinations, or a deposited layer of CsI_2:Tl. In both cases the emitted light is in the 500–600 nm wavelength range. The PIN photodiode at each pixel consists of a bottom metal electrode, a thin p^+ a-Si layer, an intrinsic (undoped) a-Si layer of about 0.5 µm, a thin n^+ a-Si layer, and a transparent ITO top metal electrode. This PIN photodiode is reverse-biased to about 5 V. When the visible photons create electron-hole pairs in the intrinsic a-Si i-layer, they are collected at the electrodes and reduce the voltage across the photodiode capacitance.

Each time a row is selected in the matrix, the readout TFT transfers the charge from the photodiode capacitance to the data line and resets the voltage across the photodiode capacitance to its original value of 5 V. The readout charge is proportional to the number of X-ray photons absorbed in the scintillator and to the number of visible photons absorbed in the PIN diode.

Figure 9.8 shows the voltage versus time on the pixel node between the PIN diode and the readout TFT, for the case of several different continuous illumination levels. Each photon absorbed in the intrinsic a-Si layer of the PIN diode creates an electron-hole pair. The electrons and holes are separated by the electric field generated by the reverse bias on the diode and collected at the terminals. The resulting photocurrent discharges the

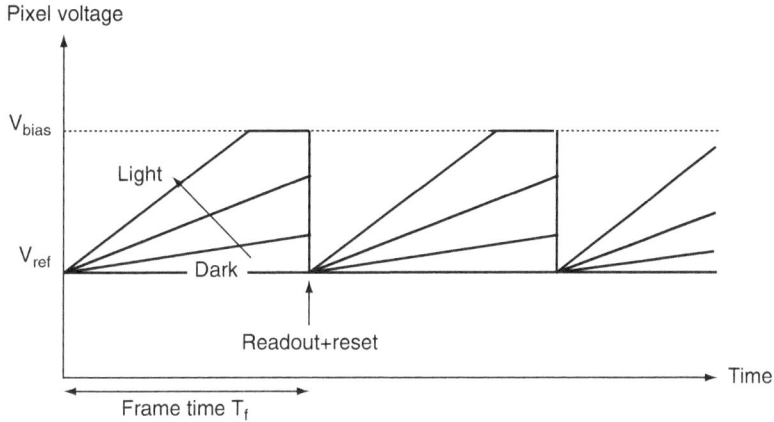

Figure 9.8: Voltage on a pixel node of indirect sensor versus time at different light levels.

diode capacitance and reduces the reverse bias across the diode. The voltage on the pixel node rises toward V_{bias} during the frame time.

Each time the pixel is read out by switching ON the readout TFTs, the pixel node is reset to the charge amplifier reference voltage V_{ref}. The charge needed to reset the pixel node flows through the data line and is converted to a voltage in the charge amplifier. Since the pixel node cannot rise beyond the common bias voltage V_{bias}, saturation of the signal occurs at high illumination levels. The maximum readout charge Q_{sat} (the saturation charge or full well capacity) is given by

$$Q_{sat} = C_{diode}(V_{ref} - V_{bias}), \tag{9.1}$$

where C_{diode} is the intrinsic diode capacitance (which functions as storage capacitor), V_{ref} is the reference voltage of the charge amplifiers, and V_{bias} is the bias voltage on the photodiodes (see Fig. 9.7).

Like the direct conversion sensor, the photodiode sensor operates in the integration mode (i.e., the total charge resulting from light exposure during the entire frame time is detected). The output voltage at the charge amplifier is

$$V_{out} = \frac{\int_{t_0}^{t_0 + T_f} I_{ph}(t)\, dt}{C_{fb}}, \tag{9.2}$$

where $I_{ph}(t)$ is the photocurrent in the diode, T_f is the frame time for readout, t_0+T_f is the readout time, and C_{fb} is the feedback capacitance in the charge amplifier. The maximum output voltage at the charge amplifier, at saturation, is given by

$$V_{out\,max} = \frac{Q_{sat}}{C_{fb}} = \frac{C_{diode}(V_{ref} - V_{bias})}{C_{fb}}. \qquad (9.3)$$

Other aspects of the operation are similar to that of direct detectors described in the previous section.

Since Gd_2O_2S:Tb screens are scattering the X-ray-induced light in multiple directions, the resolution is not as good as in direct detectors. The intensity profile is Gaussian, similar to the emission profile of a CRT shown in Fig. 5.19 in Chapter 5. In direct detectors, on the other hand, the signal profile is more square, which has given them some advantage in high-resolution applications (e.g., in mammography). To improve resolution in indirect detectors, different converter layers have been adopted, notably CsI_2:Tl. This material can be deposited on top of the TFT/photodiode array with a columnar growth technique, as shown in Fig. 9.9. The columns in the CsI_2:Tl act as light pipes, preventing significant scattering of the visible photons. By containing most of the X-ray-generated visible photons within the columns, a better resolution is achieved.

Indirect conversion X-ray sensors have been successfully applied in general radiography, in mammography (using CsI_2:Tl), and fluoroscopy [4,5,6,7]. Although their resolution is considered somewhat lower than that of direct sensors, their sensitivity can be higher, depending on the choice of the converter. This is attributed to the high gain possible during the conversion from X-ray photons to visible photons.

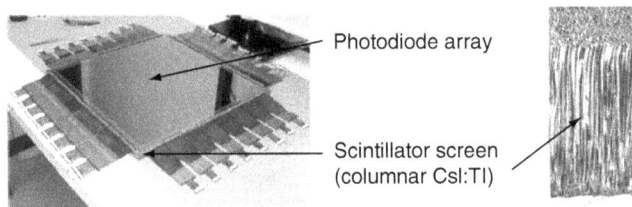

— Photodiode array

— Scintillator screen (columnar CsI:Tl)

Figure 9.9: Photograph of indirect sensor panel with scan and readout driver chips (left) and microphotograph of columnar structure in a CsI:Tl scintillator (right).

9.4   Applications of Flat Panel X-Ray Sensors

The three major applications of flat panel X-ray sensors with some of their requirements are listed in Table 9.1. General radiography is used for bones and joints as well as for soft tissues, the latter sometimes with contrast agents such as barium sulfate for gastrointestinal imaging. Mammography is also concerned with soft tissues and, particularly, with the detection of very fine calcifications.

Both radiography and mammography take snapshots. Fluoroscopy, on the other hand, is real-time continuous X-ray imaging. An example is angiography, the imaging of blood vessels after injection of an iodine compound into an artery to improve contrast. Fluoroscopy is a particularly challenging application because of the very low dose and the fast readout rate. This makes the signal-to-noise ratio much lower than in the other two applications.

Dental X-ray images are also increasingly acquired with digital detectors. Since dental detector sizes are much smaller, this market has been dominated by CCDs and CMOS image sensors, both with scintillator converters.

The use of digital detectors for medical imaging eliminates the need for large storage rooms to archive X-ray films and allows the easy transfer of image files to different locations through the Internet or local area networks. It also makes image processing and computer-aided diagnostics easier.

The digital images are acquired at 10, 12, or 14 bits of gray scale depth, and one single image file can give information on soft tissue as well as on bone structures, depending on whether the 10 most significant or 10 least significant bits of a 14-bit image are displayed. On the other hand, film radiography requires two or three different film exposures to obtain information on both bone structure and soft tissue.

The detectors usually have at least 1000×1000 pixels (sometimes more than 2000×2000). The large dynamic range reduces the need for re-exams when the patient is overexposed

Table 9.1:   Major X-ray modalities with some of their requirements

	General radiography	Mammography	Fluoroscopy
Detector size	14 in. × 17 in. or 17 in. × 17 in.	7 in. × 10 in.	12 in. × 12 in.
Pixel size	~ 140 µm	60–100 µm	200–400 µm
X-ray energy range	30–120 keV	20 keV	30–120 keV
Dynamic range	12 bits	12 bits	6 bits
Readout time	1–5 sec	1–5 sec	~ 30 msec

or underexposed. Immediate readout also eliminates uncomfortable waiting periods for a new exposure, while the film is developed. Flat panel detectors with the right choice of X-ray converter tend to be more sensitive than film, which lowers the X-ray dose to the patient.

State-of-the-art detectors with optimized signal-to-noise ratios may operate close to the quantum limit (i.e., they can detect a single X-ray photon per pixel and as few as 1000 electrons of charge on the pixel). It is interesting to note that in this application of active matrix arrays, more pixel defects are usually tolerated than in AMLCDs. The reason for this is the possibility of removing a limited number of pixel defects by image processing (e.g., by nearest neighbor interpolation).

The manufacturing volume of X-ray sensors is several orders of magnitude lower than that of AMLCDs. This, combined with the additional system cost for electronics and software, makes the cost of a complete system several hundreds of thousands of dollars. High acquisition cost has thus far limited the market penetration to larger hospitals, while smaller facilities continue to employ film-based systems with a much lower price tag.

For some medical examinations, X-ray imaging has been receiving increasing competition from other modalities such as ultrasound and magnetic resonance imaging, neither of which uses ionizing radiation and are therefore less harmful after long or frequent exposures. For many procedures, however, X-ray imaging remains the most cost-effective and sometimes the only choice.

Flat panel detectors can, in principle, be used for document imaging as well. In this application they have not been able to gain a foothold because of the low cost of existing solutions based on CCDs and linear detectors in flat bed scanners. This is related to the ease with which visible light can be redirected with lenses and other optical elements. For X-rays there are no simple solutions to focus on a small area, which explains the need for and the success of full-field flat panel sensors in X-ray imaging.

References

1. http://www.anrad.com
2. http://www.hologic.com
3. W. den Boer, S. Aggas, T. Gu, C.B. Qiu, and S.V. Thomsen, "Similarities Between TFT Arrays for Direct Conversion X-Ray Sensors and High Aperture AMLCDs," *SID 1998 Digest*, pp. 371–374 (1998).
4. http://www.gehealthcare.com/inen/rad/xr/radio/index.html
5. http://www.trixell.com
6. http://www.dpix.com
7. http://www.varian.com/xray

Index

www.ingramcontent.com/pod-product-compliance
Lightning Source LLC
Chambersburg PA
CBHW061404210326
41598CB00035B/6096

* 9 7 8 0 7 5 0 6 7 8 1 3 1 *